Oxford Medical Publications

Diagnosis of Defective Colour Vision

Diagnosis of Defective Colour Vision

Jennifer Birch

Senior Lecturer in Clinical Optometry
The City University
London

Oxford New York Tokyo
OXFORD UNIVERSITY PRESS
1993

Oxford University Press, Walton Street, Oxford OX2 6DP
Oxford New York Toronto
Delhi Bombay Calcutta Madras Karachi
Kuala Lumpur Singapore Hong Kong Tokyo
Nairobi Dar es Salaam Cape Town
Melbourne Auckland Madrid
and associated companies in
Berlin Ibadan

Oxford is a trade mark of Oxford University Press

Published in the United States
by Oxford University Press Inc., New York

A catalogue record for this book is available from the British Library

Library of Congress Cataloging in Publication Data
Birch, Jennifer.
Diagnosis of defective colour vision / Jennifer Birch.
(Oxford medical publications)
Includes bibliographical references and index.
1. Color blindness–Diagnosis. I. Title. II. Series.
[DNLM: 1. Color Perception. 2. Color Vision Defects–diagnosis.
WW 150 B617d]
RE921.B57 1993 617.7′59–dc20 92-48986
ISBN 0-19-261870-9 (hbk.) ISBN 0-19-262388-5 (pbk.)

Typeset by Graphicraft Typesetters Ltd, Hong Kong
Printed in Hong Kong

Preface

Defective red–green colour vision is inherited in an X-linked manner and occurs in 8 per cent of men and 0.4 per cent of women. Poor perception of green is more common than that of red. Defective blue vision is rare and affects both men and women equally. Colour deficiency is caused by abnormal cone photopigments and people with different types of colour deficiency have different practical difficulties. However, all colour deficient people see fewer colours in the environment and confuse colours that look different to the rest of the population. Occasionally defective colour vision is acquired by people with previously normal colour vision due to ocular or intracranial pathology, or due to prolonged use of some prescribed drugs.

Defective colour vision fascinated scientists in the nineteenth century especially those who, like John Dalton, were colour deficient themselves. Psychophysical methods of investigation were devised and these gave rise to the design of clinical colour vision tests which are the forerunners of those in use today.

The characteristics of different types of colour deficiency are fully described in this book. The design and optimum use of clinical colour vision tests for screening, grading the severity, and classifying the type of colour deficiency are explained in detail. These include pseudoisochromatic plates such as the Ishihara test, Farnsworth–Munsell hue discrimination tests, vocational lantern tests, and colour matching tests such as anomaloscopes. Test batteries which can be used with young children or as the basis for giving occupational advice are recommended and specific examples given. Careers which require normal colour vision or in which colour deficiency is a handicap are listed.

Colour vision examinations are made by many different people including doctors, optometrists, health visitors, nurses, and teachers, all of whom will find invaluable practical information in this book. Colour deficient people will also find much of interest if only to appreciate why they see the way they do and to convince them that other people, and occasionally some animals, have exactly the same colour vision problems.

London J.B.
July 1992

Contents

Plates

1. Colour in the environment

The beautiful diversity of colour in the natural world is not merely for our enjoyment. Flowers and pollinators have evolved as a community of species in which pigments and colour vision are closely related. Flowers compete for pollinators and have developed different colours to attract birds, bats, butterflies, or moths. Contrasting pigment colours may be so exactly matched to the colour vision of a particular species as to attract none other. This reduces pollen transfer and helps to stabilize each plant species. Insects have broad spectral sensitivity and visit many different types of flower. They can see well into the ultraviolet part of the spectrum, and plants use this as an important advertising signal. Honey bees differentiate colours and can learn to look for a particular colour in their search for food.

In a complex visual environment the ability to recognize different coloured objects helps animals move around safely as well as to find food. Animals use colour to communicate. Surface pigmentation is often regulated by seasonal hormones, and readiness to mate is shown by a colour change. During courtship many animals display their most colourful attributes and the recognition of distinctive colour patterns helps to identify a partner. The importance of colour recognition is also shown by the development of protective colouring. Camouflaged animals have the same body colour as their surroundings and colour patches are structured to break up the outline of the animal and conceal its shape. A camouflaged animal is very difficult to see as long as it remains stationary. The visual system is specially adapted to detect changes in pattern, and endangered animals either move very quickly or so slowly that pattern disruption is barely perceptible. Some predators are camouflaged. Reticulated pythons are inconspicuous in the light and shade of the forest canopy, and the tiger's stripes conceal it in dry grassland.

Some animals use colour in more subtle ways intended either to advertise or to deceive. Poisonous insects are brightly coloured and easy to catch, but birds learn that they are unpleasant to eat and leave them alone. Other nonpoisonous species living in the same habitat have developed identical colouration in order to benefit from the same response. Bony fish have excellent colour vision and the most brilliantly coloured species are found in sunlit tropical reefs. In this crowded environment colours have to be particularly vivid in order to register. At extreme depths, some fish provide their own light and colour by photoluminescence. These glowing colours are produced by colonies of bacteria contained in pouches on the surface of the fish. The bacteria continuously emit light as a by-product of their metabolism, but the fish is able to 'switch them off' by raising screening tissue or by restricting the blood supply to the area. These fish move around in shoals and use coloured light signals, instead of body colour, to identify each other. Even in this unique habitat predators, known as 'angler fish', have developed the capacity to produce identical light signals to deceive and lure their prey.

People use colour in many similar ways to those observed in the animal kingdom. Coloured uniforms identify members of a group and act as a form of camouflage against personal recognition by outsiders. 'Eye-catching' colours display prominence and artificial hair colour or cosmetics can be interpreted as either a form of advertisement or as an intention to deceive. However, people also appreciate aesthetic colour qualities. The colours we choose to decorate the environment reflect our personalities and manipulate our moods.

In modern society, colour is an important means of conveying messages and communicating information. Colour codes can be denotative or connotative. Denotative colour coding is widespread in business and industry. No specific meaning is attached to each colour, but colour recognition reduces search times and improves performance in sorting tasks. Connotative codes convey specific information such as electrical resistance, chemical composition, or rights of passage in transport systems.

Coloured light signals are an efficient way to achieve long distance visual signalling. The eye has a limited capacity to resolve spatial detail and both alphanumeric and shaped signs are unsuitable. Coloured light signals are very little affected by adverse weather conditions and can be seen in daylight and at night. Three or sometimes five different colours are employed. Red, green, and yellow signals are used to control road and rail transport. Red, green, white, and yellow are employed for maritime and aviation signals, and blue is used additionally for mandatory instructions. Connotative codes have gained widespread acceptance in society: red is accepted as meaning stop, danger or hazard; yellow is associated with warning or caution, and green with safety or go. The actual colours used for these codes are defined by international standards organizations so that they can be identified correctly anywhere in the world without the need for language interpretation.

Complex colour coding is an integral part of the electrical and electronics components industries. The basic colour code for resistors and capacitors contains ten colours corresponding to the ten digits from 0 to 9: these are black, brown, red, orange, yellow, green, blue, violet, grey, and white. The colours are usually grouped in three bands to give the value of the component. Further bands may be added to indicate the tolerance or voltage: the additional colours silver, gold, and pink are used for this. The background colour of most small resistors is either blue–green or brown and the majority have four superimposed bands of colour. Multicore electrical cables contain twelve colours. These are the same ten colours as for resistors but with pink and cyan added. If more than twelve different categories are required, cables are marked with twisting stripes of two colours.

Colour codes are used for pipelines, chemical containers and cylinders containing experimental gases. These containers also have written labels or number codes to assist identification and to indicate emergency procedures in the event of accidents. Most experimental gas cylinders are pink and identifying colour bands are placed around the neck of the cylinder. Red and yellow bands indicate inflammability and toxicity respectively. Blue, white, and black codes differentiate common anaesthetic gases.

Colour is used extensively in education. Children learn colour names at an early age and reading books make extensive use of colour coding. Some reading books assign a particular colour to groups of letters which have the same pronunciation or give a distinctive colour to new words which are being added to the vocabulary. Colour coded building blocks are employed to teach arithmetic. Ten separate colours are needed to represent the digits from 0 to 9. In more elaborate mathematical systems, the idea of colour mixing may be introduced to demonstrate multiplication or sets. Older pupils encounter colour coding in chemistry and in geographical maps.

The widespread use of coloured visual display units has greatly increased the scope of denotative colour coding in school as well as in the scientific and business world. Operators working with monochrome screens find that it is visually more comfortable to work with a dark grey background and white lettering than the reverse. High luminance backgrounds can lead to visual fatigue and prolonged after-images may be noticed if the unit is used for long periods. There have been a number of studies to determine the optimum use of colour to enhance visual performance. These investigations have determined the speed of knowledge acquisition with various colour combinations at different luminance contrast values. In general, the results emphasize the importance of maintaining good luminance contrast between the text and the background colour of the screen. The background colour should always be darker than the colours chosen for the text. The actual colours available depend on the design capabilities of the system. High quality colour monitors provide an enormous colour range, but the use of too many colours can actually slow down information retrieval. A maximum of five colours is recommended for data processing. However five colours may not be enough for elaborate information systems. Nine colours are currently used in some cockpit display panels for electronic flight instrument systems, and ten colours are required for digital mathematical codes. Possible colour combinations for codes containing 2–9 colours are shown in Table 1.1. (Hunt 1987.) These represent colour combinations preferred by people with normal colour vision if black cannot be used to convey information. Many of these combinations rely on red/green discrimination which is difficult for colour deficient people unless there are lightness differences between the colours.

Information shown on TV monitors can be subjected to colour enhancement, often described as false colour, to highlight information which would not otherwise be available. Some aerial photographs and medical images are treated in this way. Photographs taken with infra-red sensitive

Table 1.1 Recommended colours for codes containing 2–9 different categories

Number of colours in code	2	3	3	4	5	5	6	7	7	8	8	9
White			X			X	X	X	X	X	X	X
Cyan (turquoise)			X	X	X	X	X	X	X	X	X	X
Green	X	X		X	X	X	X	X	X	X	X	X
Yellow										X	X	X
Amber		X		X	X	X	X	X	X			
Orange			X							X	X	X
Red	X	X		X	X	X	X	X	X	X	X	X
Mauve								X		X		X
Magenta (pink)					X		X		X		X	
Purple								X		X		X
Blue									X		X	X

Reproduced from Hunt 1987.

film can have areas of different temperature displayed in colour. For example, enhanced satellite photographs can demonstrate features such as the direction of ocean currents or the effect of urbanization and deforestation on the environment. In computerized tomography, the demonstration of temperature gradients in human tissue provides information on the integrity of the vascular system and can identify tumours.

The widespread decorative use of colour in the environment means that colours need to be measured and compared in a number of manufacturing industries. Comparative colour judgements have to be made in industries where accurate colour matching is required. The comparison of small colour differences is usually necessary over a wide colour range and under a variety of lighting conditions. This type of colour quality assurance is essential in textile and paint manufacture and is fundamental to the printing and photographic industries. Although colour matching tasks can be controlled by computer, colour measuring devices have not yet achieved the sensitivity of the human eye and a final visual assessment is necessary.

Aesthetic judgements of colour are made professionally in design industries. Colour schemes are selected for their evocative qualities and to correspond with ideas of sophistication or good taste. Colour selection may be considered as the hallmark of good design in the graphic arts and in the fashion and advertising industries.

The production of colour

Light is part of general electromagnetic radiation. The human eye can detect wavelengths between about 380 and 780 nanometres (nm); where a nanometre is defined as 10^{-9} m. These wavelengths define the limits of the visible spectrum and people with normal colour vision can see about 150 different colours in this range. Sources of radiation which emit broadly within the visible range produce white light. Colour is produced by light sources which emit a limited spectral range and by filters which transmit a portion of the incident white light. Colour is also produced by pigments which reflect light selectively. The following colour names are associated with different wavelengths.

Below 380 nm	ultraviolet
380–450 nm	violet
450–490 nm	blue
490–560 nm	green
560–590 nm	yellow
590–630 nm	orange
630–780 nm	red
above 780 nm	infra-red

Incandescent light sources contain substances which emit radiation when heated to high temperatures. The amount of energy released varies with temperature. When a perfectly radiating substance, or black body, is heated the emitted light changes from reddish-white to bluish-white as the temperature is raised. Hence a method of specifying near-white colours is to refer to the equivalent colour of a perfect radiator at a particular temperature. A high colour temperature refers to a bluish-white light. Tungsten lamps emit some radiation in all visible wavelengths, but the strongest emission is in the long-wave part of the spectrum giving the resulting illumination a reddish appearance and a colour temperature of 2856. The relative spectral emission of tungsten depends on the operating temperature and changes slightly with the age of the lamp. It is therefore advisable to record the number of operating hours of any light source used for precise optical measurements, so that it can be replaced before significant changes in colour temperature occur.

Discharge lamps emit radiation when electrical energy is passed through a gas or vapour. These lamps emit line spectra superimposed on a continuous spectral distribution. The result may be highly coloured, as in sodium discharge lamps, or may have the appearance of white light. Some fluorescent light sources look white while emitting strongly in only two or three wavelength bands and the appearance of a coloured surface may alter dramatically when different 'white' lights are used for illumination (Fig. 1.1a). Only some white fluorescent lights have a balanced spectral distribution and are suitable for colour matching work. Special care is needed in selecting appropriate light sources if a natural colour appearance is needed. The colour rendering index of the lamp defines this property. A colour rendering index greater than 90 indicates good colour reproduction capabilities.

Optical lasers provide monochromatic light at extremely high energy levels. Lasers have many specialized applications including eye surgery. The argon laser is usually the instrument of choice for retinal detachment

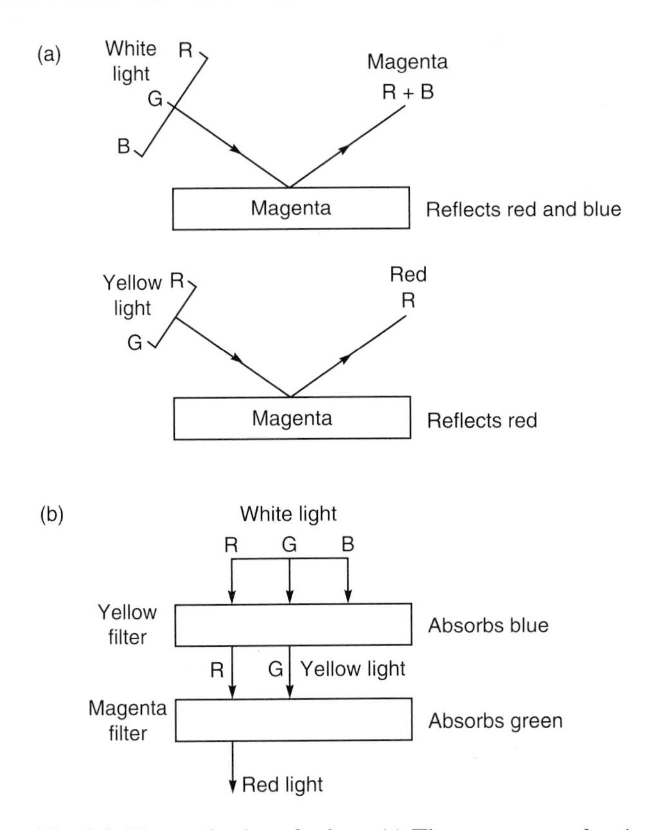

Fig. 1.1 The production of colour. (a) The appearance of a pigmented surface depends on the wavelength content of the incident light. A surface may look magenta when illuminated with white light, but red when illuminated with yellow light. (b) Transparent filters produce colour by absorbing a portion of the incident light. Placing filters in series demonstrates subtractive colour reproduction. Yellow and magenta filters absorb blue and green wavelengths from white light and only red light is transmitted.

surgery and for the treatment of retinal vascular lesions such as in diabetic retinopathy. Argon emits light at two wavelengths; 488 nm (blue) and 514 nm (green). There are three pigments in the eye which can absorb enough of these wavelengths to produce a therapeutic burn. These are melanin in the retinal pigment epithelium, haemoglobin in vascular tissue, and xanthophyll the yellow macular pigment. It may be desirable to select the most effective wavelength for a particular type of surgery in order to benefit from different absorption characteristics and to reduce light scatter in the optic media. Light scatter is maximal for short wavelengths and light damage to visual cells can occur at energy levels much lower than that required to produce a therapeutic burn. Lasers which emit longer wavelengths may be preferred for treatment to the macula because absorption of argon wavelengths by the macular pigment reduces the amount of light reaching subretinal tissue. The krypton laser emits at three wavelengths; 531 nm (green–yellow), 568 nm (yellow–orange), and 647 nm (red). Still more wavelengths are available with a dye laser. Diode lasers have recently been introduced for retinal surgery and eximer lasers are used for surgery to the anterior eye. These lasers emit wavelengths outside the visible range.

Colour can be obtained subtractively using filters. The transmittance of a filter at any particular wavelength is defined as the ratio of transmitted to incident light and is expressed as a percentage. If a filter transmits an approximately equal amount of light in all visible wavelengths it is referred to as a neutral density filter. The optical density of a filter is given by the logarithm to the base 10 of the transmittance. If two or more filters are placed in succession in a beam of light the optical densities can be added together. Glass or gelatin filters absorb some colours strongly and have the same colour appearance as the broad band of wavelengths that they transmit. For example, a red glass placed in front of a white light absorbs all wavelengths except red; a green glass absorbs all wavelengths except green and so on. Superimposing a series of colour filters in front of white light demonstrates the principle of subtractive colour mixing (Fig. 1.1b). Interference filters are frequently used in optical instruments. These filters have the advantage that they transmit a very narrow band of wavelengths and thus produce very pure colours, but since the percentage transmittance is low a high energy light source is needed to produce an intense colour.

The appearance of a coloured object depends on its reflecting properties and the wavelength content of the illumination incident upon it. A red object reflects only the red portion of the incident white light as shown in Fig. 1.1a. The reflectance of shiny or glossy surfaces depends on the geometry of measurement. Standard measurements are always made at an angle of 45 degrees to the surface being illuminated. Matt surfaces are always preferred for colour vision tests composed of pigment colours so that changes in appearance due to specular reflection are avoided. Objects tend to be recognized as having the same colour whatever the illumination or viewing conditions. This phenomenon is known as colour constancy. Colour constancy occurs over a wide range of photopic (daylight) illumination levels for the same light source, but does not hold for scenes composed of small colour differences, such as pseudoisochromatic plates, if the spectral content of the illuminant is changed. For example objects which look yellow in white light look considerably redder when viewed with tungsten illumination. However, the object is still described as yellow because it's appearance is compared with other objects in the scene which have undergone similar small transformations. Colour constancy is related to colour memory. Cortical processes compensate for changes in colour appearance so that the same colour name is ascribed to familiar objects. Coloured patterns are not subject to the same association.

The transmittance or reflectance of a particular colour sample can be measured in a spectrophotometer. The measurement is made by comparing the intensity of a beam of light after it has been transmitted, or reflected, by the sample compared to a calibrated reference standard. The reference standard for reflectance measurements is a perfectly diffusing white surface such as a block of smoked magnesium oxide. Measurements are made for a series of narrow wavelength bands positioned at regular intervals in the spectrum. The spectrophotometer is usually interfaced with a computer and calculations of reflectance can be made for light sources with different colour temperatures using one set of measurements.

Colour specification—the CIE system of colour measurement

Manufacturers and designers of coloured materials need to refer to an objective colour measurement and specification system. Such a reference system was established in 1931 by the Committee Internationale de l′ Éclairage (CIE). Measurements are based on the trichromatic colour matching characteristics of a 'standard observer'. In visual colorimetry most colours can be matched by a simple addition of three primary colours. The colour being measured is then specified by the amount of each primary contained in the match. If the sum of these quantities always represents unity, only two quantities need to be specified and the resulting gamut of colours can be represented on a two dimensional graph. Addition of three primary colours, in suitable amounts, produces white and in some cases the best trichromatic match is too desaturated, or mixed with white, to match the test colour exactly. To obtain a match, an additional desaturating colour has to be added to the test colour. This artefact leads to negative amounts in some colour equations. In order to avoid negative values the CIE adopted a system of colour measurement based on theoretical rather than actual primary colours. The addition of equal amounts of these primaries produces white light. The reference white light adopted by the CIE is an equal energy source which emits a constant amount of energy at all wavelengths in the visible region. The CIE primaries or tristimulus values are designated as X, Y, and Z. The relative amounts of each tristimulus value required in a colour match are represented by lower case letters x, y, and z; x and y values are referred to as chromaticity coordinates. Colours can be specified algebraically and graphically for any luminance level and colour measurements made with real primaries can be converted to the CIE system using a simple mathematical formula. The curved triangular boundary of the standard CIE chromaticity diagram represents the spectral locus (Fig. 1.2). The pure nonspectral purples, which are obtained by an additive mixture of red and violet light, are represented on the straight line joining the extremes of the long- and short-wavelength limits. Desaturated colours are represented within this boundary and white is placed at the centre. The coordinates of different colour temperature whites are positioned along a central locus.

A fundamental property of the chromaticity diagram is that additive mixtures of two colours always lie on a straight line joining their coordinates and the exact specification depends on the relative amounts used in the match. Specification of dominant wavelength, complementary colour, and purity can be obtained graphically (Fig. 1.3). If a straight line is drawn from a coordinate representing a particular colour to that of a specified illuminant and then extended to intersect with the spectral

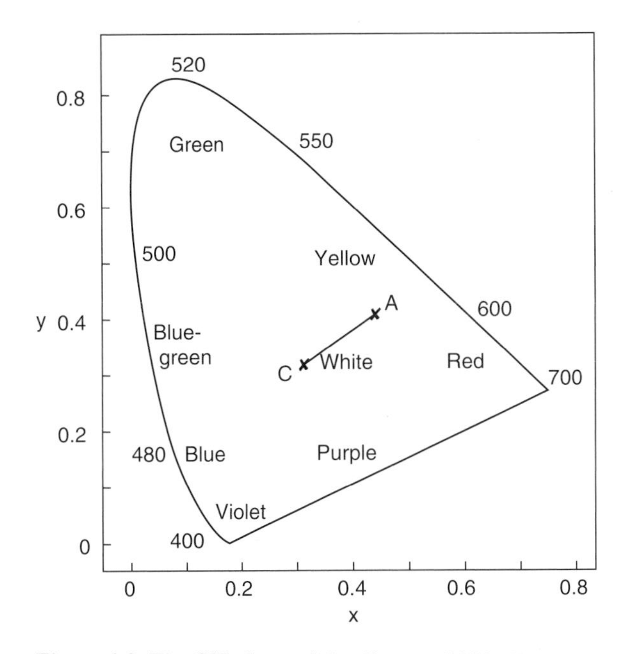

Figure 1.2 The CIE chromaticity diagram (1931). White light is specified at the centre of the diagram and spectral colours are located on the curved wavelength locus. The nonspectral purples are specified along the straight line joining the spectral limits. Desaturated colours are positioned within the locus of hues. The chromaticity coordinates of standard white light sources are:

Source A (tungsten)	$X = 0.448,\ Y = 0.407$
Source B (direct sunlight)	$X = 0.348,\ Y = 0.352$
Source C ('North' sky)	$X = 0.310,\ Y = 0.316.$

locus, the position of the intersection gives the dominant wavelength of the colour. If a different illuminant is used, the alteration in dominant wavelength shows the change in subjective appearance. Additive mixtures of complementary colours in the correct proportion produces white. If the line joining the colour coordinates with the illuminant is extended beyond white to intersect with the spectral locus and the purple line, the two points of intersection identify complementary colour pairs. Purples have no dominant wavelength but the complementary colour is a specific wavelength located on the spectral locus. If a person looks at an intense coloured light or surface for a minute or so and then at a white surface, an after image composed of the complementary colour is seen. The measured purity of the colour is equivalent to saturation or the amount of white light present in the colour. White has a purity of zero and spectral hues have a purity of unity. Colorimetric purity is shown by the relative position of the colour on the straight line joining white and the spectral locus.

The quality of natural 'white' daylight varies considerably depending on the amount of sunlight and the time of day. To ensure accurate colorimetric measurements several standard white light sources have been specified by the CIE. Illuminant A corresponds to the light provided by

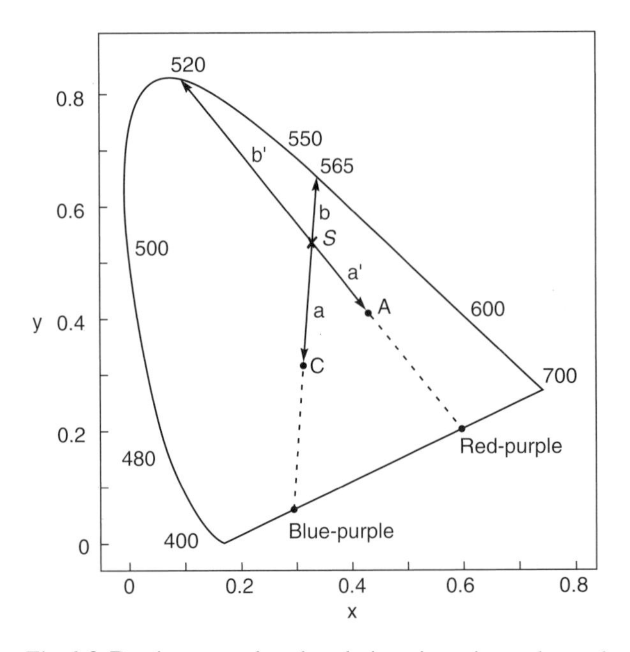

Fig. 1.3 Dominant wavelength, colorimetric purity, and complementary colours. The dominant wavelength of sample S is 565 nm when illuminated with source C and 520 nm when illuminated with source A. The complementary colour changes from a blue–purple to a red–purple when the light source is changed. Colorimetric purity is defined as the amount of white present in the sample and is obtained from the fraction: a/a + b.

a tungsten filament lamp, illuminant B to direct sunlight, and illuminant C to overcast 'north' sky light in the northern hemisphere. Illuminant C is preferred for colour matching and is a bluish-white light with a colour temperature of 6700. A slightly warmer white, defined as illuminant D or D65 (6500), is more typical of average daylight.

In the standard CIE chromaticity diagram, equal numerical differences do not represent equal perceptual differences. The ability of the standard observer to discriminate small colour differences is represented in a series of ellipses which vary in size according to the location of the reference colour in the chromaticity diagram. A chromaticity diagram with more uniform colour spacing, the UCS diagram, was approved by the CIE in 1960. The tristimulus values are designated as U, V, and W. The new chromaticity coordinates u, v, and w are derived algebraically from x, y, and z by.

$$u = \frac{4x}{-2x + 12y + 3} \tag{1.1}$$

$$v = \frac{6y}{-2x + 12y + 3} \tag{1.2}$$

$$w = \frac{3y - 3x + 1.5}{6y - x + 1.5} \tag{1.3}$$

The colour difference between specified colours can be expressed numerically. Formulae used to calculate colour differences between individual colours used in clinical colour vision tests are derived from uniform colour spacing. Several colour difference formulae have been developed and approved by the CIE over the years. These represent different attempts to produce uniform colour scaling. The first method applied to colour vision tests expressed colour differences in terms of NBS (National Bureau of Standards) units. One NBS unit is equivalent to a colour difference representing five 'least perceptable differences'. This unit has been superseded by two new formulae recommended by the CIE in 1978. These are the CIELUV formula, which is derived from the 1960 UCS diagram, and the CIELAB formula, which is derived from a further theoretical transformation designed to produce equal perceptual colour differences. The CIELAB formula is most suitable for analysing the colour differences used in colour vision tests. The colour difference (ΔE) between any two colours is calculated from eqn (1.4).

$$\Delta E^*_{ab} = [(\Delta L^*)^2 + (\Delta a^*)^2 + (\Delta b^*)^2]^{1/2} \quad (1.4)$$

where $L^* = 116\,(Y/Yn)^{1/3} - 16$
$a^* = 500\,[(X/Xn)^{1/3} - (Y/Yn)^{1/3}]$
$b^* = 200\,[(Y/Yn)^{1/3} - (Z/Zn)^{1/3}]$
and $X/Xn,\ Y/Yn,\ Z/Zn > 0.01$
X, Y, Z are the tristimulus values of the colours
Xn, Yn, Zn are the tristimulus values of the illuminant
(for standard illuminant C, $Xn = 98.0705$, $Yn = 100$,
$Zn = 118.225$)

Colour specification—the Munsell system

Most people prefer to look at a sample chart to help them choose colours rather than quote a set of mathematical coordinates. Many different charts are available. For example, one of the most comprehensive chart systems is that used by the Royal Horticultural Society to record flower colours. National colour standards, or reference colours, have been developed for use in industry and these are consulted to ensure that articles made by different manufacturing processes match in colour.

The Munsell colour appearance system was first developed in 1905 and has an international reputation for accuracy and reliability. The revised *Munsell Book of Colour* contains over 1200 colour samples in both matt and glossy finishes. The samples are arranged in uniform colour difference steps representing equal perceptual differences. Munsell colours have been correlated with the CIE system so that Munsell notations can be converted to colorimetric measurements and vice versa. The colour appearance of each Munsell colour is described in terms of hue, value, and chroma. These attributes are represented in a three dimensional structure or in the pages of a book (Fig. 1.4). Hue is defined as the quality expressed by colour names and is equivalent to wavelength or dominant wavelength. There are five principal hues in the Munsell

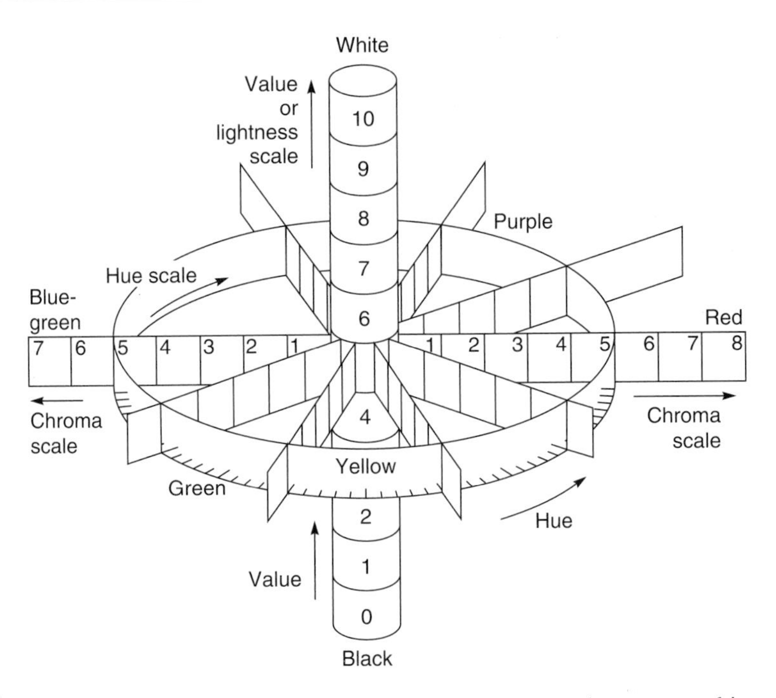

Fig. 1.4 The Munsell colour system. Colour appearance can be represented in a three dimensional diagram. Hue (dominant wavelength) is defined by a circle or by pages in a book. The value (lightness) scale is represented vertically and the chroma scale, which is equivalent to the amount of colour present, is placed radially or horizontally. Munsell samples are always specified numerically in terms of hue, value/chroma.

system blue (B), green (G), yellow (Y), red (R), and purple (P) and five intermediate hues making ten in all. Intermediate hues have composite names such as blue–green (BG), red–purple (RP) and so on. Each hue specification is divided numerically into four subgroups giving a total of forty hues. Subgroups are numbered either 2.5, 5, 7.5, or 10. Hue is represented either by a circle or by the pages of a book. The number of value and chroma steps varies with the hue. Munsell value refers to the degree of lightness or the amount of light reflected by the sample and is equivalent to photometric luminance. In the Munsell notation, black has a value of zero, and a white fully reflecting surface a value of ten. Value is represented in a vertical scale. Chroma is defined as the amount of colour present and corresponds to colorimetric purity. Chroma is similar to, but not exactly the same as, saturation. The saturation of a colour is defined as colourfulness in proportion to lightness, and differences in saturation can involve changes in both value and chroma. A white surface has a chroma of zero and a highly saturated surface a chroma of sixteen or more depending on the luminosity of the colour. Chroma is represented either radially or horizontally. Individual colours are always specified in the Munsell system by listing the hue, value/chroma such as 2.5 BG 5/4 or 10 RP 8/6. The notation system is best illustrated by a three dimensional construction but it is more practical to store the colour samples in the pages of a book with each page containing all the samples for a particular hue. Specified Munsell samples are frequently used in colour vision experiments and in diagnostic tests for defective colour vision.

Further reading

Hunt, R.W.G. (1987). *The reproduction of colour* (4th edn). Fountain Press, London.

2. Colour vision theories

One of the first explanations of human colour vision was made by Sir Isaac Newton in the seventeenth century. Newton was a member of Trinity College, Cambridge, from 1661 to 1701. His interest in optics began in about 1663 when he started work on the construction and performance of telescopes. In 1666 he travelled to Stourbridge Fair near Cambridge in order to buy prisms to examine 'the celebrated phenomenon of colours'. Newton began by placing a prism in front of a hole in a shutter in a darkened room. A beam of sunlight then passed through the prism forming a spectrum on a sheet of white paper. Newton was convinced that the spectrum was a property of sunlight and not introduced by the prism as suggested by some of his contemporaries. He confirmed this theory by passing light through a series of prisms, to show that no more colours could be produced, and by showing that white light could be reconstituted by superimposing spectral colours. Newton observed that the spectrum was composed of a continuous succession of colours. However he was persuaded by his assistant to identify seven principal hues. These colours: violet, indigo, blue, green, yellow, orange, and red, are still referred to as the seven colours of the rainbow. In selecting seven principal hues, Newton was merely conforming to current scientific practice which followed the mathematical teaching of Pythagoras by ascribing seven categories to all natural phenomena. In fact the visible spectrum can be adequately described using six colour names. The inclusion of indigo as a separate spectral colour is not justified since a gradual transition from blue to violet occurs with no distinct colour in between.

Newton considered that the perception of colour was similar to the appreciation of musical notes and suggested that both were propagated by vibrations in the ether. In a much quoted passage in *Opticks* published in 1704, Newton states that 'For the rays to speak properly are not coloured. In them there is nothing else than a certain power and disposition to stir up a sensation of this or that colour.' Newton suggested that the sensation of colour vision was produced from sympathetic vibrations set in motion 'on the bottom of the eye' by the arriving light.

Trichromatic theory

Newton's ideas about colour vision were revived and expanded by Thomas Young in his lectures to the Royal Society in 1801. Thomas Young (1773–1829) qualified in medicine at St Bartholomews' Hospital in London and although his principal occupation remained that of a physician he was also a versatile experimental scientist. While still a medical student he embarked on a detailed study of the mechanism of ocular accommodation, the success of which led to his election as a fellow of the Royal Society at the age of 21. In proposing his undulating, or wave, theory of light and his trichromatic theory of colour vision, Young was careful to claim historical authenticity by referring to Newton's publications. Young found it impossible to believe that the number of different retinal receptors could be equal to the number of spectral colours seen, and proposed that each point of the retina 'could be put in motion more or less forcibly by three primary colours'. Young conceived that colour vision was derived from a sort of colour mixing process and that there were three types of receptor, or nerves, which responded maximally to red, yellow, and blue light. However only seven months after his original lecture, the experiments of William Wollaston led Young to change his primary colours to red, green, and violet. Wollaston achieved a greater degree of prismatic dispersion than hitherto and observed that there was only a very narrow band of yellow in the spectrum and that this was produced by overlapping red and green light. Young agreed that since yellow could be obtained by colour mixture it could not be a primary colour. Green could not be obtained in this way and so must be the third primary colour. It is now more usual to refer to long, medium, and short wavelength primaries than to colour names.

Young's trichromatic theory was ignored for about fifty years until it was taken up almost simultaneously by the Scottish phycisist James Clerk Maxwell (1831–79) and the German physiologist Herman von Helmholtz (1821–94). Experimental confirmation of trichromacy was obtained by Maxwell in a series of elegant experiments using a revolving disc or spinning top. Maxwell's rooms were in the same Trinity College square in Cambridge as those previously occupied by Newton, except that Maxwell's rooms faced north whereas Newton's had faced south. Maxwell could not use direct sunlight in his experiments and had to content himself with colour mixtures derived from painted colours. A small coloured disc, representing the colour to be matched, was mounted at the centre of a spinning top and three larger coloured discs representing the three primary colours were mounted behind it. The larger discs could be interleaved so that different angular sectors of each colour were exposed until the correct match for the test colour was observed as the top rotated. The proportion of each primary colour required for the match was obtained from a protractor scale. Maxwell went on to study the effect of different types of illumination on colour matches and to devise a colour box for experimenting with spectral colours. Maxwell's colour box may be considered as a prototype for modern colorimetry and the mathematical notation which he used to calculate his colour mixtures was the earliest form of colour specification.

Helmholtz also experimented with colour mixtures but at first rejected Young's concept of three retinal receptors because he found that some trichromatic mixtures were too desaturated to match a spectral colour exactly. Helmholtz later realized that this could be explained by the three retinal receptors having overlapping spectral sensitivities. This realization enabled him to accept trichromacy unreservedly. Helmholtz' analysis of experimental colour mixture data was of fundamental importance and thereafter trichromacy became known as the Young–Helmholtz theory of colour vision. However other colour vision phenomena were less easy to explain by trichromatic theory and the position of yellow as a unique psychological sensation was difficult to reconcile.

Opponent colour Theory

In 1870 the German physiologist, Edwald Hering (1834–1918) proposed an alternative colour vision theory based on subjective colour appearance. The Hering theory proposed four primary colours, red, green, yellow, and blue arranged in opponent pairs (red opponent to green and yellow opponent to blue) with a third light/dark or luminance mechanism in which white is opponent to black. Hering's theory stressed the importance of yellow as a primary sensation and more easily explained complementary colours and coloured after images. The concept of three neural channels was retained.

Zone theories

Both the Young–Helmholtz and the Hering theories could be supported experimentally and it seemed that neither could be absolutely correct. A theoretical impasse followed. The issue was finally resolved by reviving the theory, proposed by Donders in 1881, that colour vision was processed in a series of zones in the visual pathway; trichromacy could occur at a particular level and opponency at another. This concept forms the basis of modern colour vision theories and is supported by a variety of experimental techniques.

At the receptor level, vision is trichromatic and mediated by three different types of cone (the Young–Helmholtz theory). Electrical signals from these three types of receptor are processed in the neural layers of the retina and, at the level of the ganglion cells, two opponent colour channels and a luminance channel can be detected (the Hering theory) (Fig. 2.1). The luminance channel derives its main input from the long- and medium-wave sensitive cones and has a very small input from the short-wave sensitive cones. Colour signals are processed in an opponent manner from the retinal ganglion cells to the primary visual cortex but the detailed characterisics of spatial organization may vary from level to level in the visual pathway. The type of observation and the parameters of a particular test determine whether trichromacy or opponency are needed to interpret the experimental results.

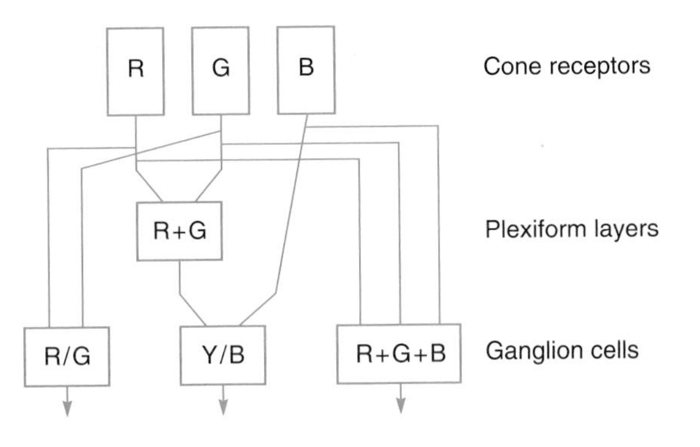

Fig. 2.1 Retinal structure and colour vision processing. Colour vision is derived from three classes of cone photopigments and is trichromatic at the receptor level (a Young–Helmholtz system). Electrical signals from the three types of cone are coded in the plexiform layers of the retina so that three opponent electrical signals can be detected at the ganglion cell level (a Hering opponent pairs system). Horizontal and Müller cells connect receptors and ganglion cells laterally within the plexiform layers and give rise to receptor fields and colour adaptation effects.

Evolution of colour vision

An evolutionary theory of colour vision was proposed by Elizabeth Ladd-Franklin in 1892. Ladd-Franklin's studies of mammalian colour vision led her to suppose that red/green discrimination had evolved fairly recently and was superimposed upon a much more ancient form of dichromatic colour vision, which enabled animals to distinguish long and short wavelengths. Only some nocturnal animals have a single cone photopigment and only primates are routinely trichromatic. Most mammals, including dogs and cats, are dichromats combining a medium-wave sensitive photopigment and a short-wave photopigment. The wavelengths of maximum sensitivity of these two pigments vary in different species. The peak sensitivity of the medium-wave pigment is positioned at discrete intervals in the spectrum between 490 and 570 nm and that of the short-wave pigment between 440 and 360 nm. Recent studies of mammalian photopigment genes suggest that the step-like changes in the medium-wave photopigment can be explained by straightforward reorganization of the amino acids in the gene molecule and that slightly different photopigments can arise from a single nucleotide substitution. Some animals, such as rodents, have a short-wave pigment with peak sensitivity in the ultraviolet. The gene specifying rhodopsin, the photopigment present in rod receptors which serves vision at very low levels of illumination, has been localized on chromosome 3 and the gene specifying the short-wave cone photopigment on chromosome 7. The amino-acid sequence of the gene specifying the short-wave cone pigment is as different from those of the long and middle-wave pigments as it is from the sequence encoding rhodopsin. In a paired comparison, the short-wave pigment shares about 42 per cent of its amino acids with rhodopsin, 43 per cent with the long-wave cone pigment, and 44 per cent with the medium-wave pigment.

This suggests that the short-wave cone pigment originated a very long time ago. In contrast, the long and middle-wave photopigments, located together in a tandem array on the X chromosome, have 96 per cent of their amino acid sequences in common. The position and similarity of these two genes suggests that they developed from a single gene much more recently in genetic terms.

From the estimated rate at which photopigments diverge, the short-wave sensitive cones may be judged to have developed about five hundred million years ago. This provided the animal with a dichromatic form of colour vision which enabled warm (reddish) and cool (bluish) objects to be distinguished. Some colour deficient people retain this form of colour vision. The short-wave receptors contribute very little to the perception of spatial detail and constitute only a few per cent of all cone cells in primate species. Electrical signals from short-wave cones appear to be processed separately in the visual pathway and in man this mechanism is found to be particularly vunerable to acquired damage.

The second subsystem of primate colour vision gives fine hue discrimination in the red–green part of the spectrum and is of considerable advantage to fructiferous animals in their search for ripe berries and fruit. Colour deficient people have great difficulty performing this task. It is possible that red–green discrimination came about by duplication, and reassembly, of the gene encoding the ancestral medium-wave cone pigment due to misalignment during meiosis. This theory is supported by the fact that there is no separate neural channel for the system. Red–green chromatic information is processed in the same neural channel as that serving spatial discrimination.

References

Mollon, J.D. (1989). The uses and origins of primate colour vision. *Journal of Experimental Biology*, **146**, 21–38.

Further reading

Sherman, P.D. (1981). *Colour vision in the nineteenth century*. Adam Hilger, Bristol.

3. Normal colour vision

The human retina contains two classes of photoreceptor. Cones, which are responsible for daytime vision, and rods which function at night. These two types of cell differ anatomically. Rods consist of transverse laminar discs which are formed continuously at the base of the cell outer segment. Each disc moves upward in time and is eventually shed. Discarded disc material is removed by the phagocitic action of the retinal pigment epithelium. Cone laminations are continuous with the receptor wall and the renewal process involves a patchwork replacement mechanism. Colour vision is present only when cones are stimulated in daylight (photopic vision) or in twilight (mesopic vision) and is absent at night (scotopic vision). There are approximately 7 million cones and 120 million rods in the human retina. The two types of cell are not uniformly distributed. The central retina surrounding the fixation point, the fovea, is essentially rod free and pure cone responses can be obtained by limiting psychophysical measurements to a central 2° field of view. Rods have their maximum density 5° from the fovea and both types of cell diminish in number towards the retinal periphery. The central area of the retina is known as the macula or macula lutea. Rods and cones differ in their sensitivity to the direction of the incident light. Cone sensitivity is greatly reduced if light enters the eye at the pupil margin and obliquely incident monochromatic light appears to change in both hue and saturation. This phenomenon is attributed to the wave-guide properties of the cone receptors and is known as the Stiles–Crawford effect. Rods do not show directional sensitivity.

The electrical activity of the retina can be demonstrated in the electroretinogram (ERG) and in the electroculogram (EOG). The ERG has two principal components; a negative a-wave which represents the activity of the receptor layer and a positive b-wave which originates in the neural layers. In certain measurement conditions, small wavelets known as oscillatory potentials can be demonstrated in the b-wave which are reduced or absent in retinal vascular lesions. Measurement of the electroretinogram is not very useful for colour vision studies as it is mainly dominated by rod responses. Under scotopic viewing conditions,

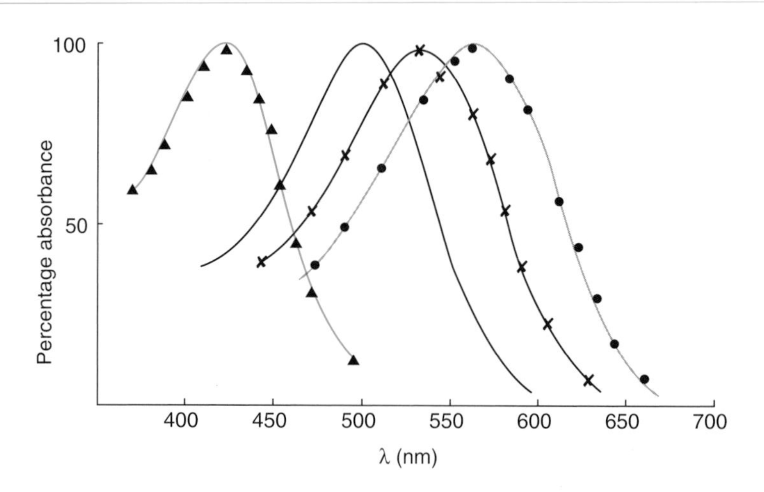

Fig. 3.1 The mean absorbance of human photopigments. The long, medium, and short-wave sensitive cones have maximum absorbance (sensitivity) at about 560 ('red'), 530 ('green'), and 420 ('blue') nm respectively. Each photopigment has broad spectral sensitivity and these overlap significantly. Δ = Short-wave sensitive cones; × = medium-wave sensitive cones; • = long-wave sensitive cones; — = rod receptors. (From Dartnall *et al.* 1983.)

the ERG is an important technique for showing pathological damage in the peripheral retina where rods have their highest concentration. Under photopic conditions, the ERG is generated by the retinal cones and is an indicator of changes in the central retina where cones have their highest density. The viewing conditions have to be carefully manipulated in order to isolate the three chromatic responses. Measurements of the 'blue' cone ERG have been the most successful. The EOG measures the standing potential of the eye and is derived from the normal interaction between the receptors and the pigment epithelium. An abnormal EOG is found in pathological conditions affecting these tissues.

Normal colour vision is trichromatic. All the spectral hues can be matched by an additive mixture of three primary colors taken from the long-wave (red), medium-wave (green), and short-wave (blue) parts of the spectrum. Although the three primaries are referred to as red, green, and blue, the wavelengths of peak sensitivity do not correspond exactly with these colour names. The retinal cones contain three classes of photopigment with overlapping spectral sensitivity which have maximum sensitivity in these spectral regions. Psychophysical and electrophysiological measurements, as well as analysis of light reflected from the retina, have shown that the maximum sensitivity of the long- and medium-wavelength sensitive cones is at about 560 and 530 nm respectively. The short-wave sensitive cones have peak sensitivity at about 420 nm (Fig. 3.1). These pigments have been given the names erythrolabe, chlorolabe, and cyanolabe. However, spectrophotometric recordings from individual cone cells have found clusters of photopigments with slightly different wavelength maxima, suggesting that the wavelength of maximum sensitivity is derived from the mean value of each pigment cluster. Recent studies of human photopigment genes show that small changes in wavelength of maximum sensitivity can arise from alteration of the sequence of amino acids in the gene structure. Thus small differences in photopigment sensitivities may define the normal distribution of colour matching characteristics. The three classes of cones are present in the proportion 20 red : 40 green : 1 blue in the central retina. The paucity of short-wavelength sensitive cones at the fovea leads to the phenomenon of 'small field tritanopia' and the normal retina is essentially blue-blind if measurements are made with a field size subtending $<0.5^\circ$.

The empirical basis for trichromatic colour vision was obtained from colour matching experiments before the measurement of absorption spectra from single cone cells had been achieved. Experiments showed that the three matching stimuli must be drawn from the long-, medium-, and short-wave parts of the spectrum but that several different wavelength combinations could be selected. It was assumed, however, that a particular set of colour matching functions must be correct if they correspond to the actual response mechanisms of the normal eye. Standardized colour matching functions for 2° matching fields were specified by the CIE in 1931. The examination technique involves presenting a wavelength in one half of a bipartite field, and adjusting the relative amounts of three matching wavelengths in the other half of the field until equality is reached. The measurement is then repeated at discrete wavelength intervals throughout the visible spectrum. The short-wave 'blue' sensitive photopigment does not contribute to matches for wavelengths longer than about 520 nm and only two colour-matching variables are required in the longwave portion of the spectrum. This enables people with normal colour vision to make a precise colour match between a monochromatic spectral yellow and a suitable proportion of monochromatic red and green wavelengths. This match, known as the Rayleigh match, is unaffected by preretinal absorption and is an important tool for the classification of red–green colour deficiency.

The retina is a multilayered structure. Electrical signals from the three classes of cones are transmitted and recoded in the plexiform layers so that at the ganglion cell level there are two opponent colour signals, red opponent to green, yellow opponent to blue, and an achromatic signal derived from all three receptor types (see Fig. 2.1, p. 17). There are a large number of different neurones subserving each retinal area and three discrete channels cannot be distinguished anatomically. However, the majority of ganglion cells have obvious input only from the long- and medium-wave sensitive cones. Only a proportion of these cells show strong red/green opponency. The remainder have broad spectral sensitivity and serve other visual functions as well as colour vision. About 6 per cent of ganglion cells have input from the short-wave cones. The blue cones contribute almost exclusively to the opponent channel and have very little input to the achromatic channel. Another unique feature is that blue/yellow ganglion cells always give blue 'on' and yellow 'off' responses and not the reverse, so that only the excitatory half of a blue/yellow channel can be identified electrophysiologically.

The optic nerve comprises the axons of the ganglion cells and contains approximately 1 million fibres. The small number of fibres in comparison to the number of receptors shows that electrical signals, especially from rods and peripheral cones, converge to single ganglion cells. Nerve fibres forming the achromatic system are generally large and serve large receptive fields. These fibres signal changes in illumination such as that produced by rapid motion or flicker. Colour opponent fibres are small. Red–green fibres serve small retinal areas and probably also contribute to pattern detection. The receptive fields of blue–yellow fibres are slightly larger. Very little recoding of colour signals takes place between the optic nerve and the striate cortex. There is partial crossover, or decussation, of the optic nerve fibres at the optic chiasma. About 70 per cent of these fibres then pass to intermediate bodies, the lateral geniculate nucleus, and to the striate cortex. The residual 30 per cent of optic nerve fibres pass mainly to the superior colliculus and the pretectoral region. These serve to control pupil reflexes and movement detection. Detailed vision is processed in the striate cortex and there appear to be separate areas concerned with the perception of form, colour, motion, and depth. Visual area V4 in the inferior occipital cortex is particularly associated with colour perception.

The neural structure of the retina contains horizontal as well as vertical elements and summation or adaptation effects can occur over a wide area. The response of a particular cell, or group of cells within a receptor field, is modified by the state of adaptation of the surrounding retina and this gives rise to simultaneous and successive contrast effects. The spatial organization of the retina contributes to opponent colour processing at the ganglion cell level. In general, the appearance of a small area of colour is altered towards the complementary colour of the surround. For example a small yellow object superimposed on a green background looks reddish whereas the same object placed on a red background looks greenish. The same phenomenon occurs if the eye is adapted by looking at an intense coloured light before viewing an object. Simultaneous colour contrast effects can produce delightful changes in colour appearance, which are used to great effect in modern designs and abstract paintings. Similar changes can produce unwanted variations in subjective colour appearance in some pseudoisochromatic tests for colour deficiency. Colour vision lantern tests exploit simultaneous and successive colour contrast effects.

Physiological variations in normal colour vision

The principle source of variation in normal colour vision is selective absorption of short wavelength light in the crystalline lens and in the inert yellow pigment which covers the central macular area of the retina. Lens absorption increases systematically with age and has a significant effect on hue discrimination ability and threshold short-wavelength sensitivity after fifty years of age. The reduction in sensitivity is more marked if cataract develops. The yellow macular pigment permeates the neural layers of the central retina and subtends between 5 and 10° horizontally and between 3 and 5° vertically. The optical density of the pigment varies individually and remains constant throughout life. The maximum density is equal to about 1 log unit. Macular pigment is a major source of variation in threshold blue sensitivity between 450 and 490 nm in people of all age groups and individual differences make it more difficult to detect tritan colour deficiency.

The presence of macular pigment can be observed in the subjective phenomenon known as Maxwell's spot. In 1856 Maxwell described the appearance of a dark spot in the blue–green part of the spectrum and a similar 'dahlia-like' spot was observed by Helmholtz after first closing his eyes and then looking at a violet evening sky. Coloured filters can be used to make the spot easier to see. The observer first looks steadily at a bright white surface through a neutral density filter for a about thirty seconds and then through a purple filter, of the same optical density, for about two seconds. Maxwell's spot then appears as a red annulus with a bright centre superimposed on a purple background. The structure of the spot varies individually and is related to the deposition of pigment around the fovea. Subjectively Maxwell's spot never subtends more than about 3° and fades quickly. About 20 per cent of people have no macular pigment and cannot see this phenomenon. However most people find that colour-matching fields which subtend more than 3° appear nonuniform due to the distribution of macular pigment within this area. Annular matching fields have to be used to avoid macular pigment effects in large field colour-matching experiments. If peripheral colour vision characteristics are being examined, the matching field is located at least 6° from the fovea.

Psychophysical measurements

Psychophysical experiments involve the measurement of a physical stimulus required to produce a 'criterion response' from an observer. The criteria employed are that of identity, where two stimuli appear to be exactly the same, or threshold detection in which a stimulus is either just perceived or appears to be just noticeably different from another stimulus. Psychophysical measurements are used to define both normal and defective colour vision but their clinical application is limited due to the complexity of the apparatus needed and the time taken to obtain the results.

Relative luminous efficiency

The sensitivity of the eye to different wavelengths in an equal energy spectrum is known as the relative luminous efficiency or V_λ function. The sensitivity of the luminance or achromatic system can be measured using flicker photometry at about 25 hertz (Hz). At photopic levels of intensity the resulting curve has a single maximum in the green–yellow region of the spectrum at 555 nm. Under scotopic conditions, the wavelength of

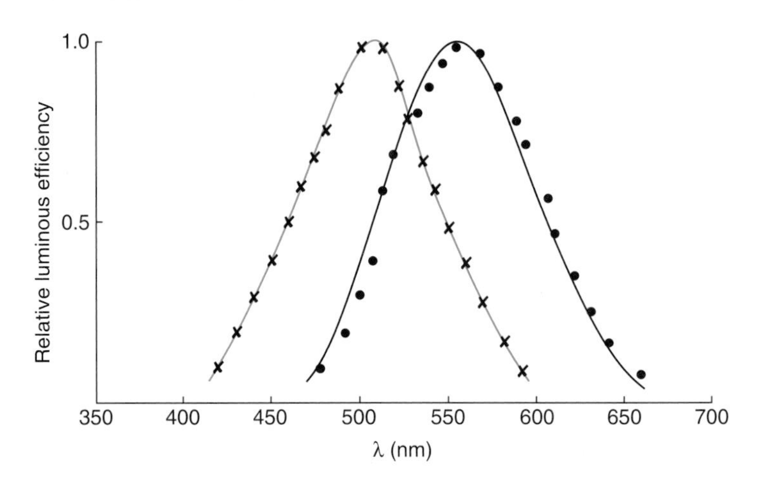

Fig. 3.2 Relative luminous efficiency obtained by flicker photometry for a 2° field. The photopic relative luminous efficiency (V_λ) curve for a 2° field modulated at about 25 Hz is derived from the three types of cone receptors. Maximum sensitivity occurs at about 555 nm. Scotopic relative luminous efficiency corresponds with the spectral sensitivity of the rod receptors and the maximum occurs at about 500 nm. The change in the wavelength of maximum relative luminous efficiency from photopic to scotopic viewing is known as the Purkinje shift.

maximum sensitivity occurs at 500 nm corresponding to the sensitivity of the rod receptors (Fig. 3.2). The change in wavelength sensitivity at reduced levels of light adaptation is known as the Purkinje shift. Spectral sensitivity can also be measured using detection thresholds. Two different methods are in current use depending on whether a white or coloured background adapting field is used. A 1° test field is superimposed on an intense adapting field subtending about 10° and is modulated at 1 Hz. The curve derived from presentations on a white background has three maxima. The peaks near 520 and 640 nm are due to the red–green opponent system and the peak at 440 nm represents the blue–yellow system (Fig. 3.3).

The use of monochromatic coloured backgrounds was pioneered by W.S. Stiles (1959) in a long series of experiments using test fields at different luminance levels. A total of seven different mechanisms, called π mechanisms, were determined. π Mechanisms are derived from complex interactions in the visual pathway and it is more convenient to investigate both normal and defective colour vision using high intensity adapting fields containing broad wavelength bands. An intense yellow background is used to adapt the long and middle wave sensitive cones so that the resulting spectral sensitivity is dominated by the short wavelength system. Spectral sensitivities obtained on magenta and blue–green backgrounds are dominated by the middle and long-wave cones respectively. The three resulting curves give a clearer demonstration of the three components of normal colour vision. These components are often referred

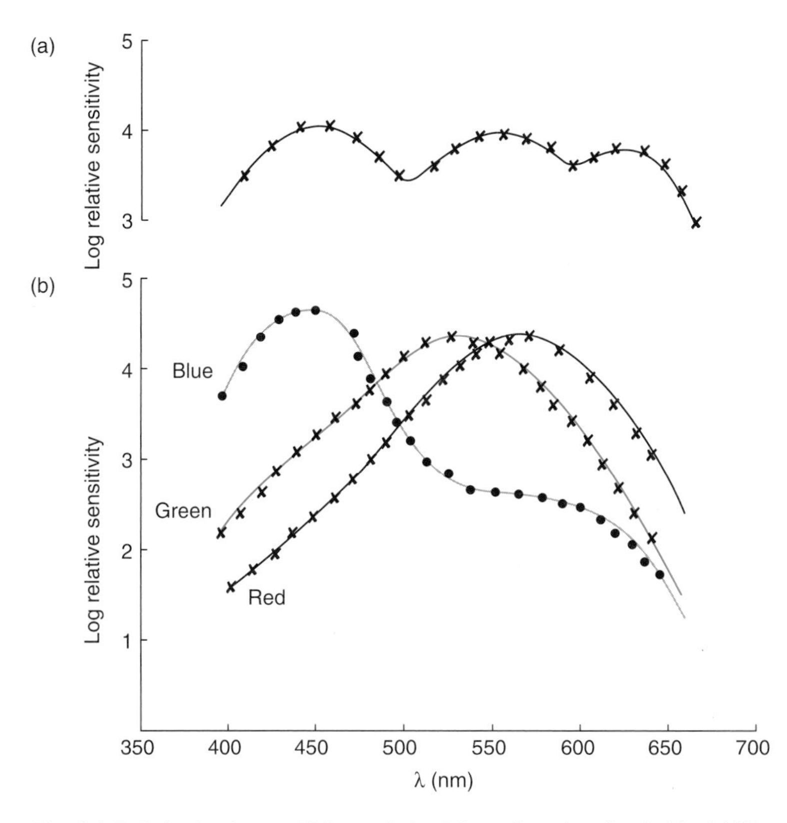

Fig. 3.3 Relative luminous efficiency derived from detection thresholds. (a) The threshold spectral sensitivity for a 1° test field modulated at 1 Hz superimposed on an intense white adapting field subtending 11° shows three maxima derived from the three colour vision mechanisms. (b) If separate yellow, magenta, and blue–green adapting fields are used, the short-wave (blue), medium-wave (green), and long-wave (red) colour vision mechanisms can be isolated (Wald–Marré mechanisms).

to as 'Wald mechanisms' or 'Wald–Marré mechanisms'. The short wavelength system has only a very small input into the luminance channel measured by flicker photometry and detection thresholds are much more useful for isolating the short-wave component of normal vision. The technique is an important tool for analysing both congenital and acquired colour deficiency which involve this system (Fig. 3.3).

Wavelength discrimination

Wavelength discrimination capacity is defined by the smallest change in wavelength which can be detected at a particular wavelength value. This is measured by introducing a wavelength into one half of a 2° bipartite field and initially the same wavelength into the other half of the field. The wavelength in the second half of the field of view is then altered until a colour difference (which is independent of a luminance difference) is

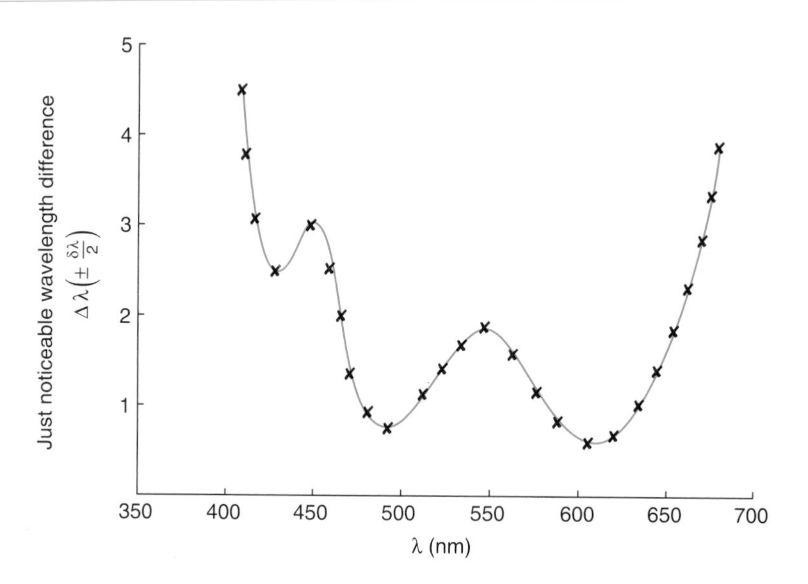

Fig. 3.4 Wavelength discrimination. The ability to detect small changes in wavelength varies throughout the spectrum. The curve for a normal trichromat for a 2° bipartite field has three minima. These occur at wavelengths where two photopigments are stimulated and the dominant sensitivity is changing from one photopigment to the other.

detected. The resulting graph is plotted with the increment step $\Delta\lambda$ at each wavelength equal to half the sum of the wavelength differences ($\delta\lambda$) found for shorter and longer wavelengths ($\Delta\delta = \pm\delta\lambda/2$) (Fig. 3.4). The curve has three minima which occur in spectral regions where two cone pigments are being stimulated differentially.

Saturation discrimination

Colour saturation, or purity, discrimination measures the quantity of a given wavelength that must be added to white light for the change in appearance to be detected. Normal saturation discrimination is optimum at about 570 nm. Saturation discrimination along specific isochromatic lines is exploited in some tests for colour deficiency.

Colour matching

It has been estimated that normal trichromats can differentiate about 3 million separate colours if all the possible variables of luminance, hue, and saturation are considered. Normal chromatic discrimination for nonspectral colours can be represented by a series of ellipses in the CIE chromaticity diagram (Fig. 3.5). Colours which have coordinates within each ellipse look the same to normal observers providing no luminance contrast exists. Discrimination ellipses were first measured by MacAdam (1942) and can be applied as tolerance limits for colour specification and colour reproduction.

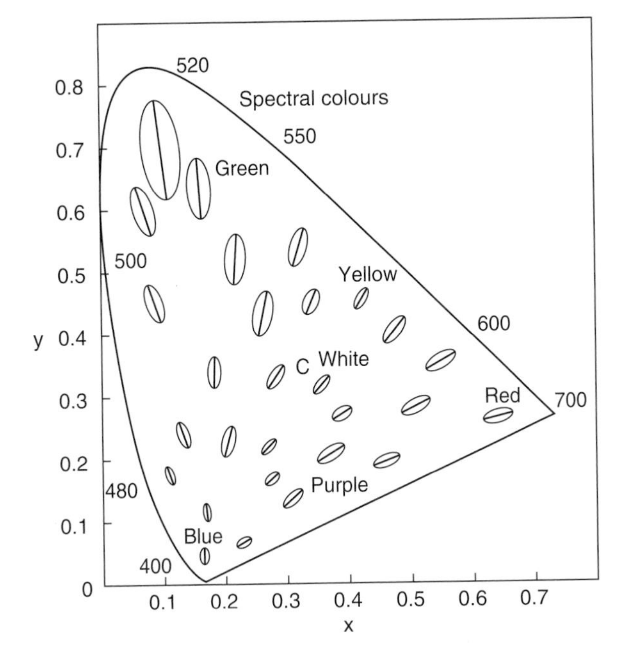

Fig. 3.5 Discrimination ellipses. Colours which have coordinates within an ellipse (shown enlarged approximately 10 times) cannot be discriminated and appear to be the same. Discrimination ellipses, known as MacAdam ellipses, represent colour matching and colour reproduction tolerances. Variations in the size of ellipses show that numerical differences in the 1931 CIE system of colour measurement do not represent equal perceptual differences.

References

Dartnall, H.J.A., Bowmaker, J.K., and Mollon, J.D. (1983). Microspectrophotometry of human photoreceptors. In *Colour vision* (ed. J.D. Mollon and L.T. Sharpe), pp. 69–80. Academic Press, London.

MacAdam, D.L. (1942). Visual sensitivities to color differences in daylight. *Journal of the Optical Society of America*, **32**, 247–74.

Stiles, W.S. (1959). Color vision: the approach through increment threshold sensitivity. *Proceedings of the National Academy of Science*, **45**, 100–14.

4. Congenital colour deficiency

The first detailed description of a person with congenital colour deficiency was made by Huddart in 1777. This was the case of Harris the shoemaker. Harris was aware that he confused colours and described the many practical difficulties which this caused. A more scientific investigation of his own colour deficiency was reported by the chemist John Dalton in 1798. Dalton made a series of observations with spectral colours and concluded that he could not see long-wavelength red light; a type of colour deficiency now known as protanopia. Dalton found that his brother had the same type of colour vision as himself and, having read about Harris, determined to investigate the possibility that some of Harris's descendants might also be affected. Dalton did not wish to leave his laboratory so he instructed a friend on how to make an examination using a selection of coloured ribbons and was delighted to discover that there were four more colour deficient people in the Harris family. This led Dalton to believe that colour deficiency might be fairly common. The ribbon test was given to fifty more people and three further cases of colour deficiency were discovered. Dalton became so interested in the subject that he corresponded with other famous scientists seeking a plausible explanation. Sir John Herschel, the Astronomer Royal, replied that he supposed that most people had three primary colour sensations and that Dalton had only two. However, Dalton rejected this correct interpretation in favour of his own idea that the vitreous humour of his eye must be coloured blue and therefore absorbed red light. Dalton's explanation was finally disproved after a *post mortem* examination, carried out at his own request, found that his optic media were clear and transparent.

Dalton's scientific papers received wide attention both in the UK and elsewhere. As a result, many more colour deficient people were discovered and the term Daltonism was used to describe them. Dalton's British colleagues found it offensive that he should be commemorated in this way and David Brewster (1829), the distinguished optical physicist suggested that the term 'colour blindness', meaning blindness to one or more colours, should be used instead.

Although both Dalton and Brewster constructed spectral colour vision tests, most case histories relied on subjective reports. The investigation of defective colour vision was therefore considerably advanced in 1837 when August Seebeck produced a systematic test consisting of 300 coloured papers which had to be sorted into groups. Seebeck showed that there were two distinct classes of colour deficiency and that differences in severity occurred in both classes. Only one of these groups was found to have reduced sensitivity to red light.

Seebeck's examination methods were similar to those used today but he lacked a proper framework for interpreting the results. This was provided by John William Strutt, the third Baron Rayleigh in 1881. Rayleigh adapted earlier spectral tests into a precise colour matching examination and discovered that some colour deficient people could be classified as dichromats and others as anomalous trichromats. Dichromats lack one of the three primary colour sensations. Anomalous trichromats have three primary sensations but one of them is abnormal. These two types of colour deficiency are sometimes referred to as 'loss' and 'alteration' systems. The discovery of different types of colour deficiency meant that 'colour blindness' no longer seemed appropriate and the terms 'colour deficiency' or 'dyschromatopsia' (meaning colours seen incorrectly) became universal. Several different terms were proposed to describe types of colour deficiency but the nomenclature suggested by von Kries (1897) was finally adopted.

In the nineteenth century, interest in colour deficiency centred on congenital red–green defects, although loss of short-wavelength (blue) colour perception was known to occur. Most of the cases of poor blue vision reported at this time appeared to be acquired in origin but appropriate terms were included in the nomenclature for congenital colour deficiency. Congenital tritanopia was not fully documented until 1952 after a group of affected individuals were identified from a survey made in the magazine *Picture Post* (Wright 1952).

Classification of colour deficiency

Defective colour vision is characterized by abnormal colour matching and colour confusions. Colours which look different to people with normal colour vision can look the same to people with defective colour vision. There is also a marked reduction in the number of separate colours that can be distinguished in the spectrum and the relative luminous efficiency of the eye is altered.

Congenital colour deficiency is caused by inherited photopigment abnormalities. These may arise in several different ways. The retina may be lacking in functional cone receptors or there may be only one or only two cone pigments instead of three. Alternatively, one of the three types of cone may contain a photopigment which differs significantly in spectral sensitivity compared to the normal pigment. The terms assigned to these different types of colour deficiency are based on the number of photopigments present and thus on the number of colour matching variables required to match all the spectral hues (Table 4.1). In addition, the terms

Table 4.1 Classification of congenital colour deficiency

Number of variables in wavelength matching	Number of cone photopigments	Type	Denomination	Hue discrimination
1	None	Monochromat (Achromat)	Typical or Rod monochromat	Absent
1	One	Monochromat (Achromat)	Atypical, incomplete, or cone monochromat	Absent
2	Two	Dichromat	Protanope Deuteranope Tritanope	Severely impaired
3	Three (one abnormal)	Anomalous trichromat	Protanomalous Deuteranomalous Tritanomalous	Continuous range of severity from severe to mildly impaired
3	Three	Normal trichromat	Normal trichromat	Optimum

protan, deutan, and tritan, from the Greek meaning first, second, and third, are used to denote which of the three photopigments is affected. The term 'tetartan', meaning fourth type, of colour deficiency was added by supporters of the Hering theory who supposed that anomalies of a 'yellow' sensitive photopigment might be found.

Monochromats

Monochromats are able to match all spectral wavelengths using one colour matching variable. Individuals with this type of colour deficiency are 'colour blind' and see only lightness differences in the environment. There are two types of deficiency. Typical or rod monochromats have no functioning cone receptors. Visual acuity is poor, within the range 6/60–6/36, and affected individuals have photophobia (aversion to bright lights) and nystagmus (unsteady fixation). Atypical or cone monochromats have a single cone type. Cone monochromatism, sometimes confusingly described as 'incomplete achromatopsia', is extremely rare and most affected individuals have short-wave 'blue' sensitive cones only. Visual acuity is usually reduced in the range 6/9–6/24 and only people with acuity less than 6/18 have photophobia and nystagmus. Monochromatism is also referred to as 'achromatopsia', meaning absence of colour vision, but the term is mainly used to describe acquired colour blindness of cortical origin. Measurement of the ERG clearly differentiates between rod and cone monochromats since a photopic response of any amplitude shows a residual cone response.

Dichromats

Dichromats have two cone photopigments instead of three and are able to match all the spectral hues using two colour matching variables. There are three types of dichromatism depending on which of the three normal pigments is missing. Protanopes lack the long-wave 'red' sensitive photopigment, deuteranopes lack the middle-wave 'green' sensitive photopigment, and tritanopes lack the short-wave 'blue' sensitive pigment. The presence of a single photopigment in the red–green spectral range has been confirmed in protanopia and deuteranopia from retinal reflection measurements. In addition, absence of the medium-wave photopigment cluster has been demonstrated from spectrophotometric recordings on an excised deuteranopic eye. Measurement of the ERG under conditions of chromatic adaptation has confirmed the absence of the relevant cone response in each type of congenital dichromatism.

Anomalous trichromats

Anomalous trichromats require three colour matching variables to match all spectral hues. Three types of colour deficiency occur depending on which photopigment has abnormal absorption characteristics. The terms used are protanomalous, deuteranomalous, and tritanomalous trichromatism referring to the presence of abnormal 'red', 'green', and 'blue' sensitive photopigments respectively. The absorption characteristics of the abnormal pigment varies individually and a continuous range of severity is found in each type of deficiency. Hue discrimination may be almost as severely affected as that of the corresponding dichromat or nearly normal. Anomalous trichromats do not accept colour matches made by normal trichromats and do not, necessarily, accept colour matches made by other anomalous trichromats in the same category.

Reflection densitometry is not sensitive enough to detect small differences in photopigment sensitivity; and whether abnormal photopigments, independent of differences in neural coding, are entirely responsible for the range of severity found in anomalous trichromatism is still under investigation. However, new data obtained from molecular genetic studies have shown that a variety of photopigments with slightly different wavelength maxima can occur.

Group terms

The group terms protan, deutan, and tritan are given to types of colour deficiency involving absence or abnormality of a single photopigment. Each term includes dichromatism and anomalous trichromatism in that category. Protan and deutan defects are described collectively as red–green colour deficiency. Red–green defects share a common mode of inheritance and similar colour confusions occur. People with severe colour deficiency may confuse bright reds and greens whereas people with slight red–green colour deficiency only confuse dark or desaturated colours.

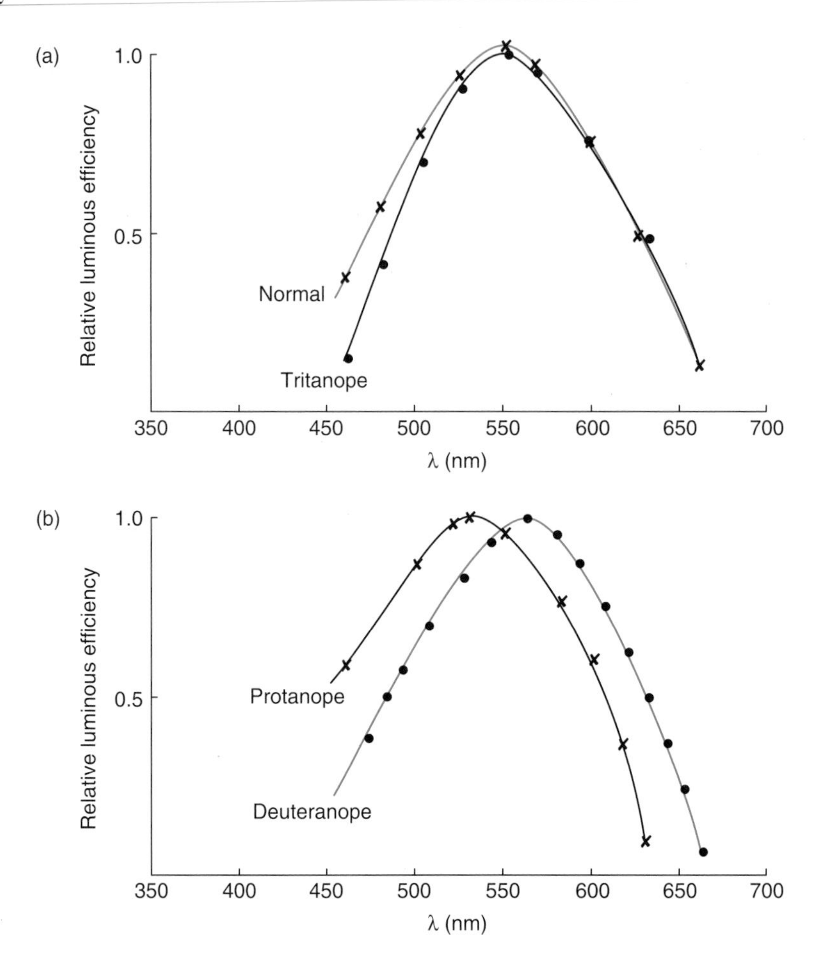

Fig. 4.1 The relative luminous efficiency of congenital dichromats obtained by flicker photometry (V_λ) for a 2° field. (a) Tritanopia. In tritanopia the relative luminous efficiency is reduced for wavelengths shorter than 540 nm compared to the normal curve, but there is no change in the wavelength of maximum spectral sensitivity. (b) Protanopia and deuteranopia. The wavelength of maximum sensitivity is at about 530 nm for protanopes and at about 565 nm for deuteranopes. In protanopia the relative luminous efficiency is reduced for wavelengths greater than 600 nm and is referred to as 'shortening of the red end of the spectrum'.

There is no loss of visual acuity in congenital protan, deutan, or tritan defects and all other visual functions are entirely normal.

Characteristics of congenital colour deficiency

Relative luminous efficiency

The wavelength of maximum relative luminous efficiency, measured by flicker photometry at 25 Hz, is different in protan and deutan defects (Fig. 4.1). The relative luminous efficiency of protans is very different from that of normal trichromats. In protanopia, maximum sensitivity occurs at about 535 nm (compared with 555 nm in normal observers) and there is a marked reduction in sensitivity above 600 nm. The absence of sensitivity for wavelengths longer than 630 nm is often described as 'shortening of the red end of the spectrum'. In deutan defects there is a much smaller shift of maximum sensitivity towards longer wavelengths. In deuteranopia, maximum relative luminous efficiency is at about 565 nm. Tritanopes have reduced sensitivity at the short-wave end of the spectrum and there is no alteration in the wavelength of maximum sensitivity. The relative luminous efficiency of anomalous trichromats is intermediate between that of normal trichromats and the corresponding dichromat. Detection thresholds measured with a white adapting field show loss of the red–green opponent system in protanopes and deuteranopes, but very careful observations have to be made in order to reveal this deficit. The technique is much more robust in showing absence or reduction of the short-wave mechanism in tritan defects (Fig. 4.2a).

Detection thresholds with coloured backgrounds (the Wald–Marré technique) show that one of the three colour mechanisms is absent in dichromatism and that one of the relevant mechanisms is altered or reduced in anomalous trichromatism (Fig. 4.2b). This technique has been used extensively in the diagnosis of congenital and acquired tritan defects and is the basis of two diagnostic tests, the TNO test and the Berkeley color threshold test.

Hue discrimination

Protanopes and deuteranopes have similar wavelength discrimination. There is a single minimum at about 495 nm and neither type of dichromat is able to detect wavelength differences in the long-wave portion of the spectrum above 520 nm. The number of separate spectral hues which can be distinguished is therefore reduced to about 30 compared to 150 in normal trichromatism. The corresponding data for anomalous trichromats show considerable observer variation. Hue discrimination characteristics are represented in a series of curves which are intermediate between the normal and dichromatic values. Some anomalous trichromats are nearly dichromatic and have only residual hue discrimination for wavelengths longer than 540 nm. In deuteranomalous trichromatism there is a shift in the position of minimal discrimination in the yellow part of the spectrum. This minimum occurs at 590 nm in normal trichromats and at 610 nm in deuteranomalous trichromats. Tritanopes cannot detect wavelength differences between 450 and 480 nm, and tritanomalous trichromats have reduced discrimination in this spectral range (Fig. 4.3).

Saturation discrimination is reduced at all wavelengths for both dichromats and anomalous trichromats, and the normal minimum at 570 nm is absent.

Colour matching

Colours perceived to be identical in dichromatic vision can be represented in a series of narrow zones in the 1931 CIE chromaticity diagram.

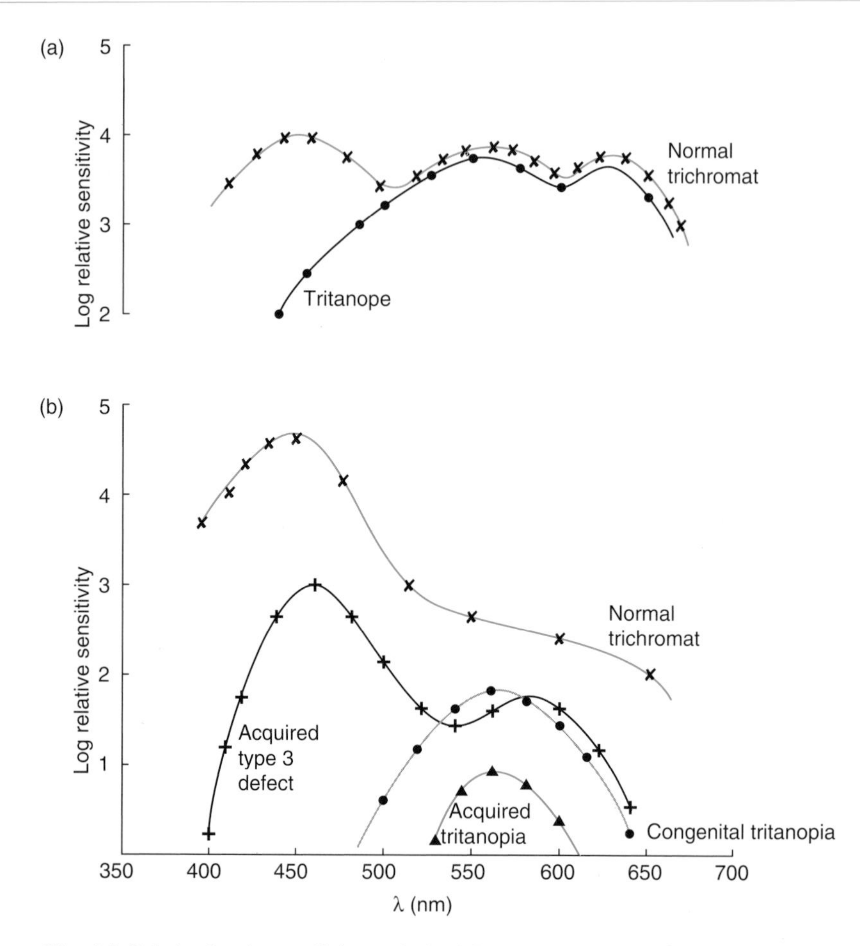

Fig. 4.2 Relative luminous efficiency derived from detection thresholds in tritan deficiency. (a) No short-wave component of the curve can be detected in congenital tritanopia for a 1° test field superimposed on a white background. (b) The characteristics of the curve obtained on a yellow adapting field depend on the intensity of the field. In most viewing conditions, the medium- and long-wave mechanisms are only partially adapted and the portion of the curve above 500 nm is due to these mechanisms. There is no spectral sensitivity at 460 nm in either congenital or acquired tritanopia. In acquired type 3 (tritan) defects found in diabetic retinopathy, the short-wave component of the curve is reduced rather than absent and the patient is effectively a tritanomalous trichromat.

Colours having coordinates within isochromatic (same colour) zones look the same and are confused by dichromats providing that no luminance contrast exists between the selected colours. Isochromatic lines represent the average data for several observers (Fig. 4.4). These lines appear to converge to a single point which, in the CIE system of measurement, is located outside the chromaticity diagram (Table 4.2). The convergence of isochromatic lines provides further evidence for the loss of a fundamental sensation in dichromatic vision. Differences in macular pigment density alters the exact location of individual convergence points. The deuteranopic

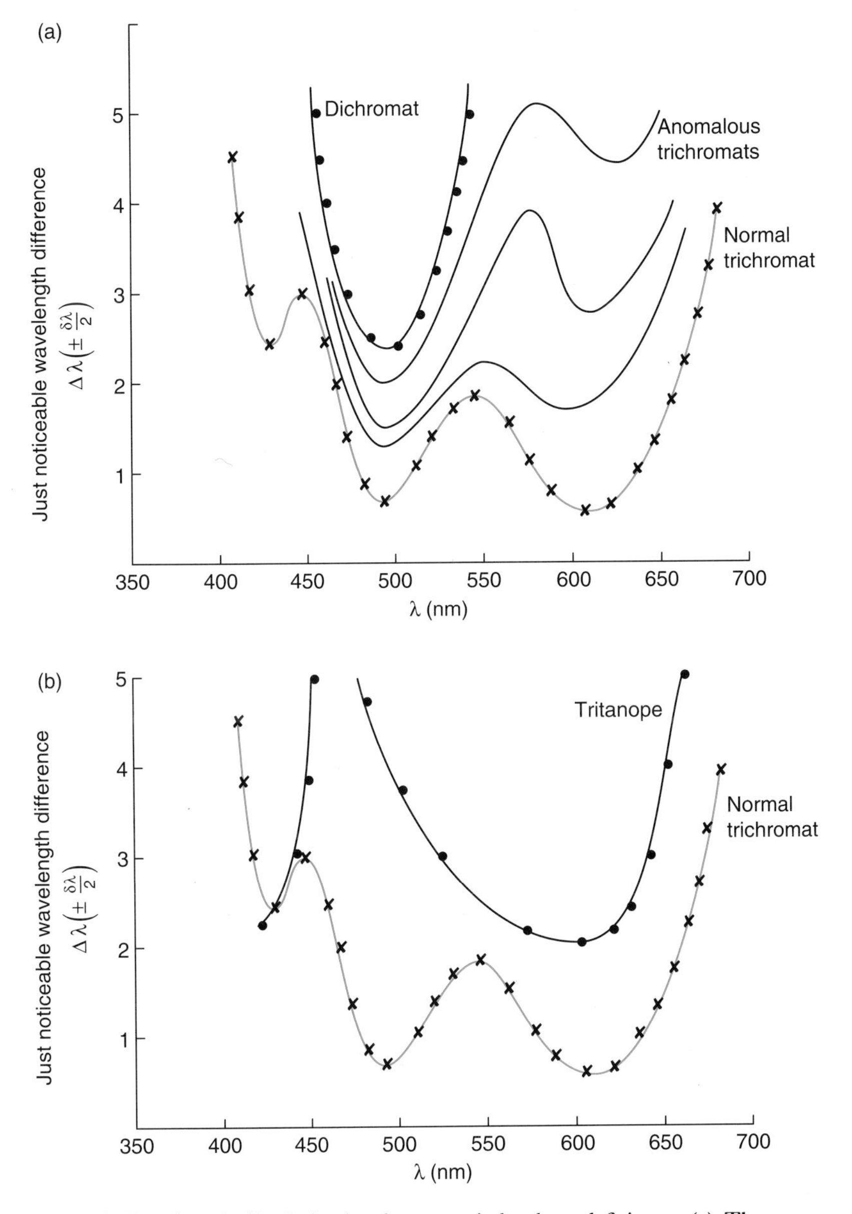

Fig. 4.3 Wavelength discrimination in congenital colour deficiency. (a) The wavelength discrimination of both protanopes and deuteranopes has a single minimum at about 495 nm and there is no discrimination of wavelengths longer than 540 nm. A series of curves, which are intermediate between that of the normal trichromat and the corresponding dichromat, is found in both protanomalous and deuteranomalous trichromatism showing that a continuum of severity occurs in these classes of colour deficiency. Some anomalous trichromats have minimal wavelength discrimination above 540 nm. (b) In tritanopia there is no wavelength discrimination between 450 and 480 nm. The resulting curve has two minima, one in the extreme violet below 440 nm and the other at about 600 nm.

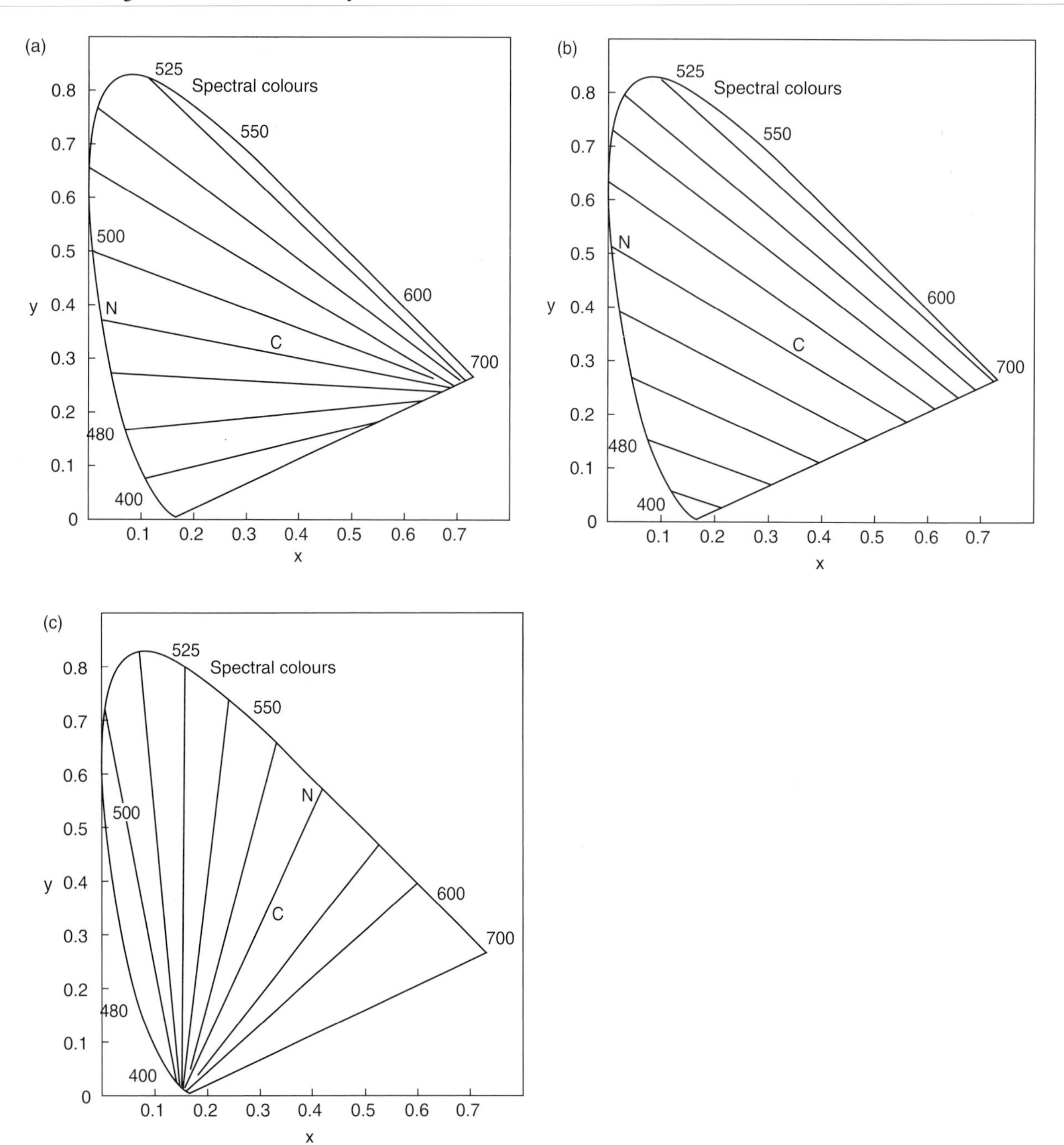

Fig. 4.4 Dichromatic isochromatic lines. Isochromatic lines represent the mean value of colour confusion zones for dichromats. Colours represented along a line look identical if no luminance contrast is present. Isochromatic lines for each of the three types of congenital dichromat converge to a point which in the CIE system of measurement lies outside the chromaticity diagram. Isochromatic lines for anomalous trichromats are similar to those of the corresponding dichromat but are shorter and do not include the full range of chromaticities. (a) Protanopia: isochromatic lines converge to the point $x = 0.75$, $y = 0.25$. Neutral points occur at approximately 494 nm (blue–green) and in the red–purple. (b) Deuteranopia: isochromatic lines converge to the point $x = 1.40$, $y = -0.40$. Neutral points occur at approximately 499 nm (green) and in the blue–purple. (c) Tritanopia: isochromatic lines converge to the point $x = 0.17$, $y = 0$. Neutral points occur at approximately 572 nm (yellow) and in the violet part of the spectrum. N shows the spectral neutral point for illuminant C.

Table 4.2 Dichromatic convergence points

	CIE chromaticity Coordinates	
	x	y
Protanopia	0.75	0.25
Deuteranopia	1.40	−0.40
Tritanopia	0.17	0.00

convergence points are particularly affected by differences in macular pigment density and fairly large variations in the calculated values are found.

Protanopes and deuteranopes are able to match blue–green wavelengths with nonspectral purples represented on the line joining the spectral limits in the chromaticity diagram. The points of intersection of the isochromatic line passing through white defines achromatic colours (colours which are confused with each other and with a neutral grey having the same lightness). For protanopes the achromatic colours are reddish-purple and blue–green, and for deuteranopes bluish–purple and green. The point of intersection with the spectral locus defines the achromatic or 'neutral point' (N) in the spectrum. For illuminant C, the average neutral point is at 494.3 nm for protanopes and at 498.4 nm for deuteranopes. However because of macular pigment differences, the range of individual values overlap and measurement of the spectral neutral point is an unreliable method for distinguishing between red–green dichromats. There is potentially a larger colour difference between the protanopic and deuteranopic neutral purples and these colours are usually chosen for protan/deutan classification designs in pseudoisochromatic tests.

Tritanopes are able to match wavelengths below 440 nm with wavelengths longer than 500 nm. Neutral points are in the yellow region of the spectrum at about 572 nm and in the extreme violet around 380 nm. The average tritanopic wavelength matches are between 440 and 500 nm, 430 and 510 nm, 420 and 530 nm, 410 and 550 nm, and 400 and 570 nm. These wavelength matches can only be demonstrated if the apparatus used for measurement provides sufficient photopic luminance in the extreme short-wave part of the visible spectrum.

Isochromatic data for anomalous trichromats are similar to those of the corresponding dichromat but do not include the complete range of chromaticities. Isochromatic lines, which define the long axis of the discrimination ellipse, are shorter in length but have the same orientation in the chromaticity diagram. The length of the isochromatic line depends on the extent of the photopigment abnormality and thus on the severity of colour deficiency. Neutral colours have the same dominant wavelength, but only desaturated colours are confused. Isochromatic colour confusions are utilized in pseudoisochromatic tests and are demonstrated in the Farnsworth D15 test. Neutral colours are used to classify protan, deutan, and tritan types of deficiency, and colour confusions between isochromatic colours with different amounts of saturation are used to grade the severity of colour deficiency.

Isochromatic lines for both protanopes and deuteranopes coincide with the straight portion of the spectral locus for wavelengths greater than 520 nm. An additive mixture of wavelengths in this spectral range always produces a fully saturated intermediate wavelength. This characteristic explains the effectiveness of the Rayleigh match in diagnosing red–green colour deficiency. Dichromats are able to match any mixture of red and green wavelengths to an intermediate yellow wavelength providing that luminance differences can be equated. The proportion of red and green required to match yellow is altered in anomalous trichromatism. Protanomalous trichromats require significantly more red in their match compared with the normal value and deuteranomalous trichromats require more green. The range of red/green values used to match yellow shows the extent of the photopigment abnormality and hence the degree of colour deficiency. Careful examination techniques identify a small number of people, extreme anomalous trichromats, who have mixed protan and deutan deficiency derived from abnormal long-wave and abnormal medium-wave photopigments. Rayleigh matches obtained with the Nagel anomaloscope for 127 unrelated consecutive male anomalous trichromats, attending a colour vision advisory clinic, are shown in Fig. 4.5. The normal mean match for the instrument is 44 on the red/green mixture scale and the standard deviation is 2 scale units. Twenty observers were found to be protanomalous trichromats, 100 were deuteranomalous trichromats and 7 were extreme anomalous trichromats who have combined protan and deutan defects. The matching ranges obtained by extreme anomalous observers encompass the normal range but do not extend to both limits of the matching scale. As expected from large surveys, deuteranomalous trichromats outnumber protanomalous trichromats by 5 to 1 and it is possible to estimate that the incidence of extreme anomalous trichromatism is about 0.3 per cent in males. Although some protanomalous and deuteranomalous trichromats have identical anomaloscope matching ranges, these data show a continuous range of severity in both types of red–green colour deficiency rather than division into a small number of subgroups. The matching ranges of deuteranomalous trichromats are clearly separated from the normal distribution of Rayleigh matches, but this separation is less clear for some protanomalous trichromats. Both protanomalous trichromats and extreme anomalous observers have reduced sensitivity to long-wavelengths, and obtain Rayleigh matches having an excess of red light at low luminance values for the yellow matching field. Anomalous trichromats who demonstrate small matching ranges on the Nagel anomaloscope have fewer practical difficulties and achieve better results on clinical colour vision tests than observers who obtain large matching ranges.

There is no equivalent straight line in the chromaticity diagram which can be exploited to diagnose tritan colour deficiency. Addition of blue and green wavelengths does not produce a unique hue and the resulting blue–green is always desaturated. The orientation of tritan isochromatic

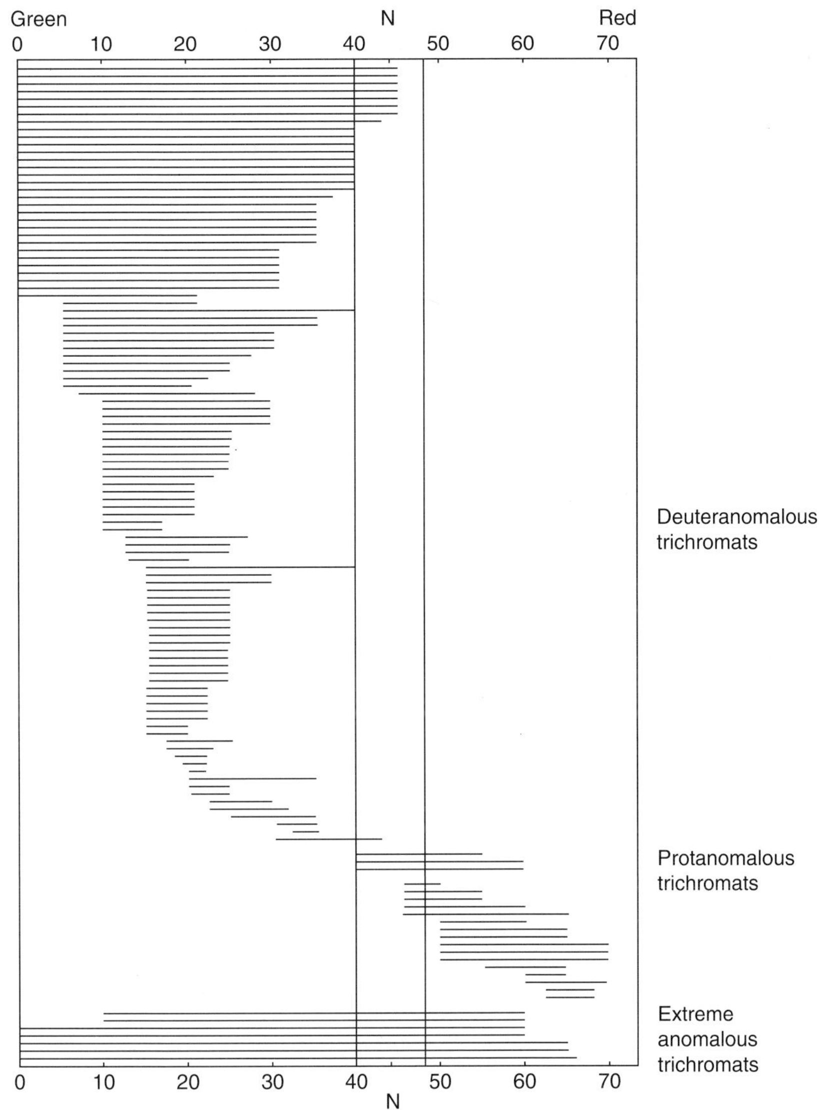

Fig. 4.5 Matching ranges for 127 anomalous trichromats obtained with the Nagel anomaloscope. The normal mean match (N) for the anomaloscope is at 44 on the red–green mixture scale (SD ± 2 scale units). One hundred observers are deuteranomalous trichromats and have matching ranges between 0 and 45 on the mixture scale. Twenty observers are protanomalous trichromats and have matching ranges between 40 and 70 on the mixture scale. The data for both protanomalous and deuteranomalous trichromats suggest a continuous range of severity rather than discrete subgroups. Seven observers are extreme anomalous trichromats and have matching ranges which include the normal matching range and extend to both the red and green sides of the mixture scale. Protanomalous trichromats and extreme anomalous trichromats have reduced threshold sensitivity to long wavelengths compared with normal trichromats.

lines in the chromaticity diagram is very similar to the long axis of normal discrimination ellipses, and isochromatic colours with fairly large colour differences are needed, in pseudoisochromatic tests, to avoid false positive results by normal trichromats with high macular pigment density. Direct wavelength matches or measurement of detection thresholds are required to confirm tritanopia.

References

Brewster, J. (1829). Account of two remarkable cases of insensitivity in the eye to particular colours. *Edinburgh Journal of Science*, **10**.

Dalton, J. (1798). Extraordinary facts relating to the vision of colours. *Memoranda of the Literary and Philosophical Society of Manchester*, **5**, 28.

Huddart, J. (1777). An account of persons who could not distinguish colours. *Philosophical Transactions of the Royal Society London*, **67**, 260–5.

van Kries J. (1987). Über Farbensysteme. *Ziets Für Psychology*, 13, 241–324.

Seebeck, A. (1837). Über den bei mancher personnen vorkcommenden Mangel or Farbesinn. *Annales van Physiology* (Leipzig), **42**, 177–233.

Strutt, J. (Lord Rayleigh), (1881). Experiments on colour. *Nature*, **25**, 64–6.

Wright, W.D. (1952). The characteristics of tritanopia. *Journal of the Optical Society of America*, **42**, 509–21.

Further reading

Wright, W.D. (1946). *Researches in normal and defective colour vision*. Henry Kimpton, London.

5. Incidence and inheritance of congenital colour vision defects

The first reports of people with defective colour vision showed that certain families were affected. Two of Harris's five brothers were colour deficient and John Dalton found that his brother had identical colour vision to himself. A year after the report of Harris's colour deficiency, another protanope, J. Scott, described defective colour vision in three generations of his own family. He found that both his father and his maternal uncle were colour deficient. His mother had normal colour vision but his brother and one of his two sisters were affected. Scott's own son and daughter had normal colour vision. Without understanding the mode of transmission, Scott had documented an X-linked inheritance with colour deficiency passing through both of his parents families resulting in a colour deficient female (Fig. 5.1). Dalton attempted to estimate the size of the problem by examining 50 people but his investigation was too small to give reliable results. A much larger survey was undertaken by George Wilson, an Edinburgh surgeon, in 1852–3. Wilson examined over a thousand soldiers and medical students using a colour naming test and concluded that 5.6 per cent of men were colour deficient. Wilson was surprised by this high prevalence rate and expressed concern about the possible practical implications of his findings.

Large scale surveys using modern screening techniques have shown that the incidence of congenital red–green deficiency is about 8 per cent in the male population and between about 0.4 and 0.5 per cent in the corresponding female population (Table 5.1). One in every 12 males and 1 in about 200 females have some form of colour deficiency.

There are many reports in the literature which appear to show that the prevalence of colour deficiency varies significantly in different ethnic groups and in different geographical areas. However, a careful review of the available data does not support this conclusion. Two main critisms can be made of these studies. Firstly, the examination techniques are often inappropriate or poorly understood by the population being

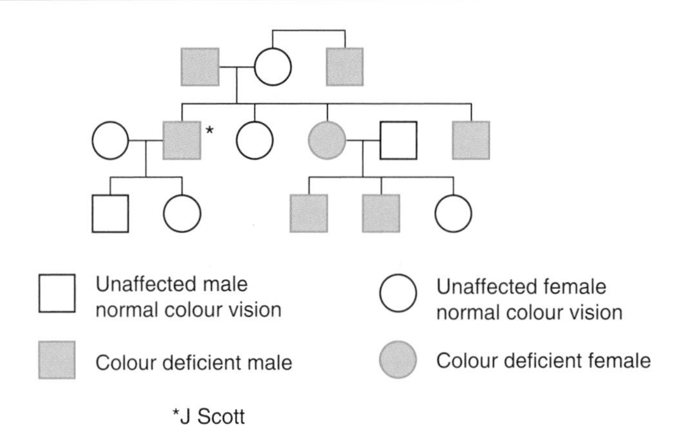

Fig. 5.1 The Scott family (1778). The Scott pedigree shows 6 colour deficient males and 1 colour deficient female in three generations.

Table 5.1 The prevalence of congenital red–green colour deficiency obtained from large-scale surveys

		Men			Women		
Country	Year	Number tested	Number colour deficient	Percentage	Number tested	Number colour deficient	Percentage
Germany	1936	6863	532	7.75	5604	20	0.36
France	1959	6635	594	8.95	6990	35	0.50
Norway	1927	9049	725	8.01	9072	40	0.44
Greece	1975	21 231	1687	7.95	8754	37	0.42
		Average percentage = 8.14			Average percentage = 0.43		

examined and secondly, the number of people surveyed is usually much too small to give reliable results. An objective yardstick is available to assess the accuracy of small surveys. The inheritance of red–green colour deficiency is X-linked and the male/female incidence ratio should conform to this pattern. For example, surveys in Japan and China in the 1930s estimated the frequency of colour deficiency to be about 5 per cent in men and between 1 and 1.5 per cent in women. In this case the prevalence in females is much too high, compared with that of the corresponding male population, for an X-linked trait and this removes confidence in the examination method. Paradoxically, Japanese pseudoisochromatic screening tests are pre-eminent in Western countries but designs containing arabic numerals may be unsuitable for rural Asian populations.

The male/female prevalence ratio shows an X-linked pattern in the survey of Australian Aborigines made by Mann and Turner in 1956. Mann and Turner examined 4500 males and 3200 females and found an incidence of 2 per cent in males and 0.03 per cent in females. These results confirm a lower incidence of colour deficiency in people of this ethnic origin only.

In large randomly mating populations the relative proportions of different genotypes remains constant from one generation to another (the Hardy–Weinberg principle), and differences in the incidence of inherited traits have to be explained in terms of population migration, mutation, or selection. The prevalence of genetically determined traits may vary from the accepted figure in communities which are isolated geographically or by social practice. In isolated communities, the gene pool is restricted and nonrandom mating results in either greater or fewer genetically determined traits compared with large unrestricted populations. Separate evolution after the formation of the Australian land mass may be responsible for the lower prevalence of colour deficiency in Australian Aborigines. Colour vision differences in the old and new world primates also suggest separate evolution following geographic isolation.

X-linked inheritance

The science of modern genetics originates from the work of Gregor Mendel in the latter half of the nineteenth century. Before Mendel conducted his experiments in plant breeding it was considered that conception involved the mingling of hereditary substances from both parents. Mendel showed that this was not so. For each physical characteristic a person possesses two hereditary factors or genes, one from each parent. If the genes are identical the person is said to be homozygous and if the genes are different the person in said to be heterozygous for that particular characteristic. In the heterozygous state one gene is dominant and the other recessive; the person manifests the characteristics of the dominant gene.

Genes are located on chromosomes, the minute thread-like structures present in the nucleus of each cell. The behaviour of chromosomes during cell division provides the mechanism for Mendelian inheritance. In man, the diploid number of chromosomes is 46. There are two types of chromosomes, those concerned with sex determination, the sex chromosomes, and the remaining 22 pairs which are referred to as autosomes. In females the two sex chromosome are similar and labelled XX; males have one X chromosome and a smaller dissimilar chromosome labelled Y. Sex-linked inheritance occurs when genes are carried on either of the sex chromosomes, the most usual pattern being X-linked. An X-linked recessive trait is one determined by a gene on the X chromosome which is not manifest in the female in the heterozygous state. Heterozygous females are often described as 'carriers' of genetic traits. A recessive gene carried on the single X chromosome is always manifest in males when there is no matching gene on the Y chromosome. Red–green colour deficiency is a prime example of X-linked recessive inheritance. The four possible combinations of X and Y chromosomes in the offspring of a colour deficient male and a normal female are shown in Fig. 5.2. Chromosome pairings occur equally by chance and in very large families there will be an equal number of all four genotypes. All the daughters inherit their fathers X

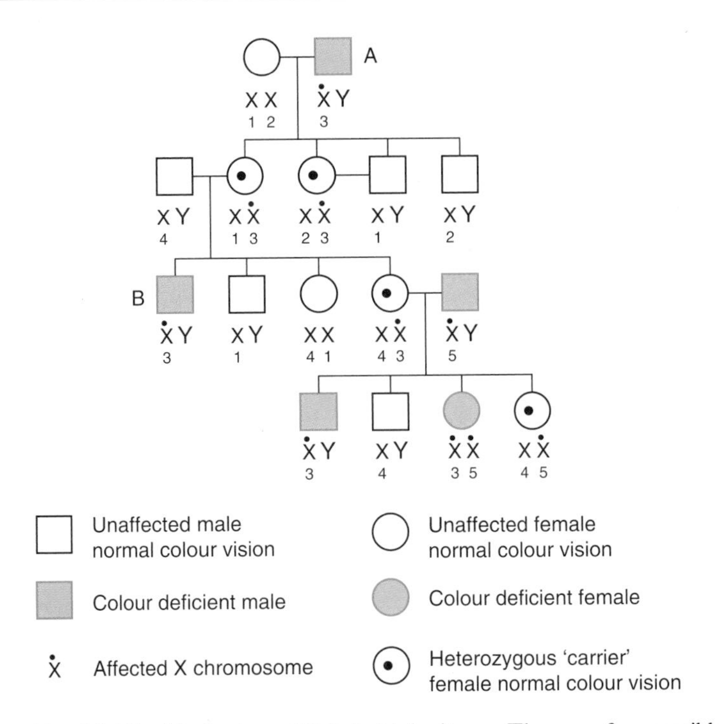

Fig. 5.2 The Mechanism of X-linked inheritance. There are four possible combinations of X and Y chromosomes inherited from the parents in each generation. Females receive X chromosomes from both parents, males receive one of their mother's X chromosomes and their father's Y chromosome. Chromosome pairings occur by chance and in very large families there will be an equal number of all four genotypes. The usual inheritance is from maternal grandfather (A) to grandson (B). (X chromosomes are numbered so that they can be identified in each generation.)

chromosome and are heterozygous 'carriers' of colour deficiency; all the sons have normal colour vision. In the case of a heterozygous woman marrying a man with normal colour vision, 50 per cent of their sons will be colour deficient and 50 per cent of their daughters heterozygous 'carriers'. That is, each offspring has a 50 per cent chance of inheriting the abnormal X chromosome from their mother. If these two events occur in sequence it follows that the most usual form of transmission in families is from maternal grandfather to grandson. The first large pedigree showing X-linked colour deficiency was published by Horner in 1876. This typical family tree traces the inheritance of deuteranopia through six generations of the same family and shows that males are affected in alternate generations (Fig. 5.3). Women are only colour deficient if they inherit a gene for colour deficiency from both parents as shown in the Scott pedigree and in Fig. 5.2.

A characteristic of X-linked inheritance is that affected family members always have the same type and severity of colour deficiency. The discovery of identical colour deficiency in families promoted the idea that

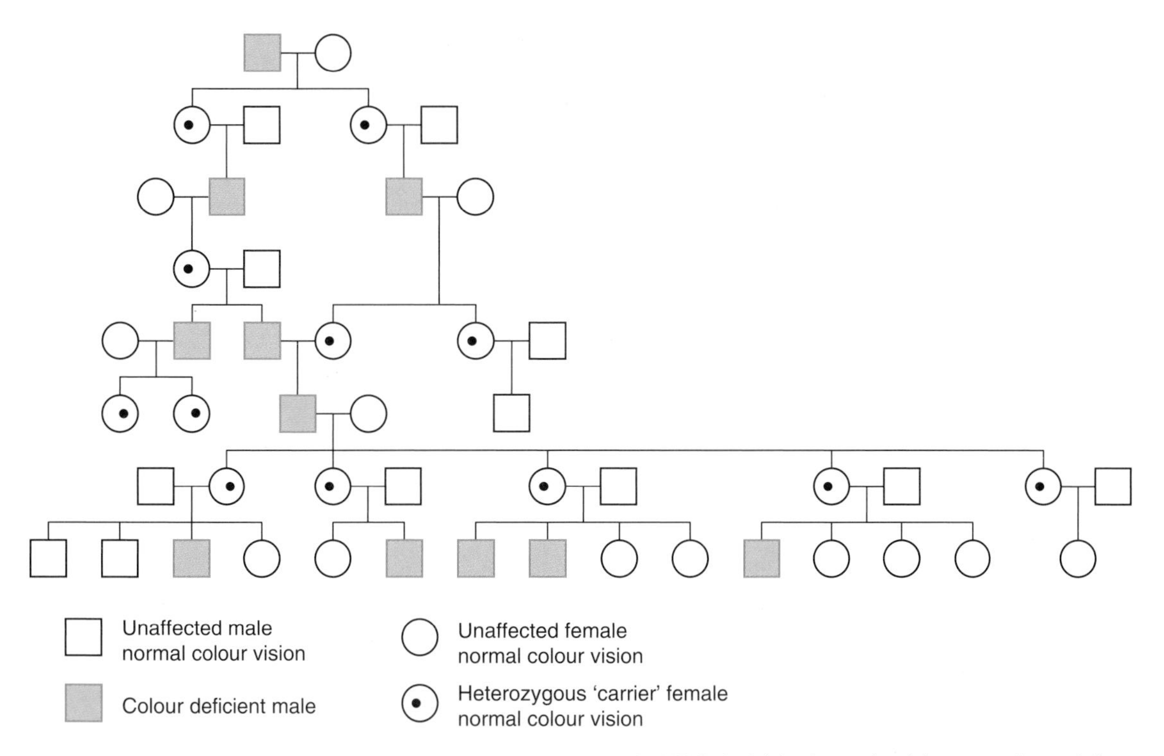

Fig. 5.3 Horner's pedigree (1876). Horner's pedigree shows typical X-linked inheritance in eight generations of the same family. Red–green colour deficient males occur in alternate generations. However, the number of colour deficient males is high, usually only half the sons of heterozygous females are affected.

there may be two specific genes for 'simple' and 'severe' anomalous trichromatism. However the presence of two distinct types of protanomalous and deuteranomalous trichromatism is not supported by hue discrimination and colour matching data gathered from a large number of unrelated individuals. These data show a continuum of severity in anomalous trichromacy in both classes of red–green colour deficiency.

The two locus hypothesis

If the probability of defective colour vision in men is 8 per cent, then the probability in women should be the square or 0.64 per cent if a single gene is involved. Large surveys show that the incidence in women is significantly lower than this figure. The discrepancy is explained if the genes for protan and deutan defects are represented on two separate, but adjacent, loci on the X chromosome and form two series of multiple alleles or alternative forms. These alleles are in rank order of dominance; normal trichromatism, anomalous trichromatism, and dichromatism. If a female is a compound heterozygote and has two genes from an allelic series she will express the dominant gene. For example, if a woman has a gene for deuteranopia on one X chromosome and a gene for deuteranomalous trichromatism on the other, she will be a deuteranomalous trichromat. Females who are mixed, or nonallelic, compound heterozygotes

Table 5.2 Prevalence of different types of red–green deficiency in men and women

Type of deficiency	Frequency in men	Frequency in women
Protanopia	1%	0.01%
Protanomalous trichromatism	1%	0.03%
Deuteranopia	1%	0.01%
Deuteranomalous trichromatism	5%	0.35%
Total	8%	0.40%

and have a gene for a protan defect on one chromosome and for a deutan defect on the other, will have normal colour vision since normal trichromatism is dominant at both loci. Some published data support this theory, but other reports show that at least some compound mixed heterozygotes have overt protan defects and that others have minimal colour deficiency shown by an abnormal Rayleigh match. These reports suggest a more complex genetic mechanism.

Screening tests are adequate for estimating the prevalence of colour deficiency, but accurate diagnosis of the type of deficiency is required in genetic studies. This is achieved either by using a spectral anomaloscope or by detailed psychophysical measurements. The different types of red–green colour deficiency do not occur with the same frequency (Table 5.2). The ratio of different types of deficiency in males is 1 protanope (P) : 1 protanomalous trichromat (PA) : 1 deuteranope (D) : 5 deuteranomalous trichromats (DA). Thus the prevalence of female compound mixed heterozygotes is 0.24 (Fig. 5.4a). If the probability of 0.64 per cent in women is reduced by this amount, the resulting figure of 0.40 per cent is in close agreement with survey data. The proportion of different types of colour deficiency in females would then be 1P : 3PA: 1D: 35DA. The large colour vision survey carried out among military service recruits in Greece in 1975 found these predicted ratios for both males and females. The agreement is less exact if a compilation is made of the most reliable European statistics. These data give a slightly higher prevalence of colour deficiency in women (0.5 per cent) due to a larger number of protanomalous trichromats than predicted. There are several reports in the literature of compound mixed heterozygotes manifesting protan defects, and if half of the predicted number manifest the condition the incidence of colour deficiency in females would be 0.45 per cent (Fig. 5.4b).

Recombination of protan and deutan genes

Recombination of genetic material during meiosis may influence the expression of colour deficiency in females. Meiosis is the process of nuclear

(a)

X	P	PA	D	DA
P	1P	1PA	—	—
PA	1PA	1PA	—	—
D	—	—	1D	5DA
DA	—	—	5DA	25DA

0.4% colour deficiency in women
in proportion 1P:3PA:1D:35DA

(b)

X	P	PA	D	DA
P	1P	1PA	—	—
PA	1PA	1PA	0.5PA	2.5PA
D	—	0.5PA	1D	5DA
DA	—	2.5PA	5DA	25DA

0.46% colour deficiency in women
in proportion 1P:9PA:1D:35DA

Fig. 5.4 The frequency of different types of red–green colour deficiency in females. Large-scale surveys show that 8% of males have red–green colour deficiency. Defects occur in the ratio 1 protanope (P) : 1 protanomalous trichromat (PA) : 1 deuteranope (D): 5 deuteranomalous trichromats (DA). The same surveys show that slightly more than 0.4% of the females have red–green defects (Table 5.2). The observed frequency of different types of colour deficiency in females may occur as follows: (a) If all compound mixed heterozygotes have normal colour vision. (b) If half of the compound mixed heterozygotes with a PA genotype manifest PA.

division which occurs when gametes are formed. Homologous or paired chromosomes may exchange parts of their genetic material during this process. In effect both chromosomes fracture and the disconnected fragments crossover and re-attach to the opposite chromosome. Crossover may result in two genes which were originally located on the same chromosome being separated, and genes which were originally on different chromosomes being brought together (Fig. 5.5). The best known example of recombination is the pedigree of Vanderdonck and Verriest (1960) in which the protanomalous daughter of a deuteranope is shown to have both deuteranopic, protanomalous, and normal sons, and protanomalous and normal daughters (Fig. 5.6). This pedigree suggests that recombination has occurred at least twice in successive generations. Careful examination of individual Rayleigh matches shows that some men as well as women have combined protan and deutan defects. In these cases recombination results in a rare type of extreme anomalous trichromatism in which both the long-wave and medium-wave photopigments are abnormal. This type of colour deficiency occurs in about 0.3 per cent of males (see Fig. 4.5).

Mosaicism

Mosaicism may be a contributary factor in producing minimal colour deficiency in female heterozygotes. In 1962 Mary Lyon drew attention to the fact that females inherit an enormous amount of duplicate genetic material on paired X chromosomes, and that some method of inactivation or suppression must occur at an early stage of development. According to

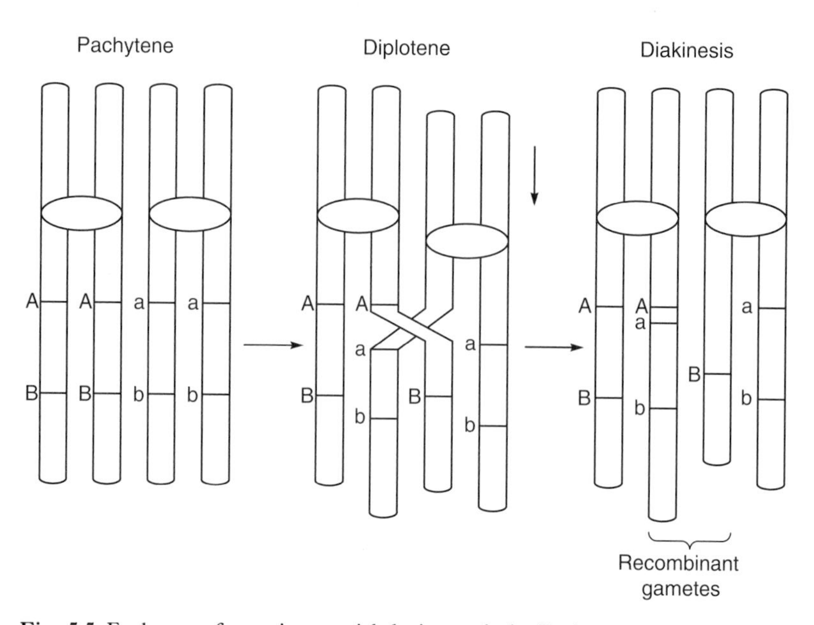

Fig. 5.5 Exchange of genetic material during meiosis. Exchange or 'crossover' of genetic material during meiosis coupled with slight nonalignment is thought to give rise to either loss of a photopigment gene or hybrid gene formation depending on the position of the breakpoint.

this hypothesis either of the two X chromosomes may be active in each somatic cell. The patchy fur pigmentation of female tortoiseshell cats, colouring never found in male cats, is an example of mosaicism. Fur colour is known to be X-linked. Female tortoiseshell cats receive a gene for brown fur on one X chromosome and one for ginger fur on the other. Suppression of the brown pigment gene results in a patch of ginger fur and vice versa. Women who are heterozygous for defective colour vision may develop a similar mosaic of retinal cone cells in which some groups of cells have normal photopigments and some do not. The proportion of abnormal cells may vary individually. This hypothesis is supported by the fact that some known heterozygotes select a red-green ratio for a Rayleigh match which is at the extreme of the normal distribution, or have an abnormally large matching range. Slight displacement of the mean match is known as Schmidt's sign. Protan heterozygotes may also have abnormal relative luminous efficiency, with the wavelength of maximum sensitivity shifted slightly towards shorter wavelengths.

Red–green colour deficiency has a high prevalence in the population and provides a 'marker' in studies of genetic linkage. The presence of colour deficiency can demonstrate gene dominance in chromosome abnormalities arising from duplicate or triplicate X chromosomes. The incidence of colour deficiency in females who have an XO chromosome complement (Turner's syndrome) is the same as in men.

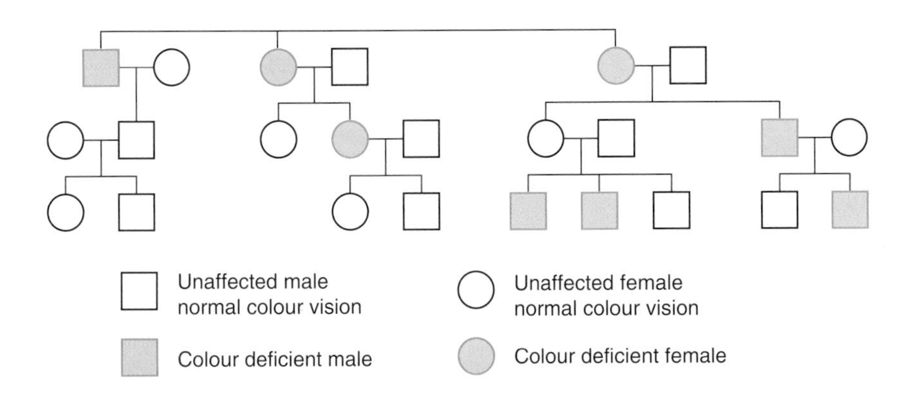

Fig. 5.6 Recombination of protan and deutan defects. The pedigree of Vanderdonck and Verriest (1960) shows a compound mixed heterozygote female manifesting protanomalous trichromatism and able to transmit both deuteranopia and protanomalous trichromatism to her sons. The fact that two other sons have normal colour vision and one daughter has protanomalous trichromatism shows that further recombination must have taken place.

Autosomal inheritance

The inheritance of tritan defects is autosomal dominant. A dominant trait is manifest in heterozygotes and affects males and females equally. Half the children of an affected parent manifest the condition (Fig. 5.7). Variable expression is a characteristic of inherited autosomal traits and differences in severity are usually found in affected family members. Some individuals are tritanopes and others tritanomalous trichromats. The survey carried out in *Picture Post* estimated the incidence of congenital tritanopia as not greater than 1 in 10 000. There is no available data for the incidence of tritanomalous trichromatism. Effective screening for tritan defects must allow for normal variations in macular pigment density and reliable clinical screening test are difficult to construct. Surveys, using the TNO test and the new City University tritan plates, have identified 2 people with tritan deficiency in a population of 450, suggesting that these defects may be much more numerous than previously supposed. It is possible for tritan defects and red–green defects to occur in the same family due to separate modes of inheritance. Concurrent tritan and deutan defects have been described in the literature.

Typical rod monochromatism is an autosomal recessive condition and consanguinity is a predisposing factor. The frequency is estimated to be about 0.003 and both men and women are equally affected. There is no reliable data for the inheritance of atypical cone monochromatism.

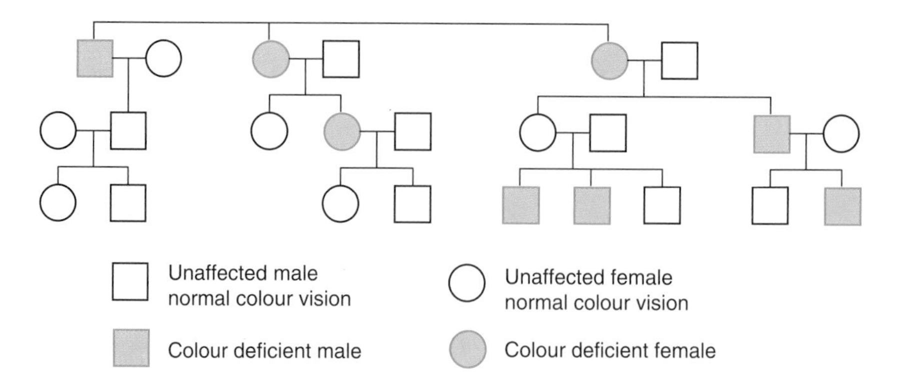

Fig. 5.7 The inheritance of tritan defects. The pedigree of Went and Pronk (1985) has 8 individuals with tritan defects in three generations and shows autosomal dominant inheritance with incomplete penetrance. Both males and females are affected and both are able to transmit the gene to half of their offspring.

Molecular genetics

The introduction of recombinant DNA technology in the early 1970s has revolutionized the science of molecular genetics. These studies emphasize that it is more appropriate to describe the inheritance of colour deficiency in terms of genes which specify photopigments rather than genes which produce different types of colour deficiency. The use of host vectors and radio-active probes enables accurate gene maps to be constructed and provides a tool for the diagnosis and control of genetic disease. The technique was first used in the study of retinal photopigments by Nathans and his co-workers in 1986. The gene encoding the rod photopigment, rhodopsin, has been located on the distal arm of chromosome 3 and that specifying the short-wave blue sensitive photopigment on chromosome 7. So far three different single nucleotide changes have been found in the short-wave photopigment gene in families with tritan defects. All three alterations are sufficient to produce a non-spectrally active photopigment. Genes specifying the normal long-wave and medium-wave photopigments have been located in a tandem array near the distal end of the long arm of the X chromosome and have almost identical amino-acid sequences. The similarity of these genes suggests that abnormal photopigments could easily be produced during meiosis, firstly by slight misalignment of two X chromosomes and then by crossover. If the two strands of DNA are slightly out of register the long-wave gene on one X chromosome could be paired with the medium-wave gene on the other. If crossover then occurs and the breakpoint lies in the duplicate spacers between the genes, one chromosome may lose a gene and the other chromosome gain one. A man who inherits an X chromosome which lacks a photopigment gene will be a dichromat. If the break occurs within a gene sequence, joining of the two fragments produces a fusion or hybrid gene (Fig. 5.5). The similarity of the two gene structures suggests that this could occur very readily. A man who inherits a fusion gene will be an anomalous trichromat. Any number of different genes, resulting in different types of anomalous trichromacy, could be formed in this way. Substitution or deletion of the hydroxyl-bearing amino acids of the gene structure appear to be involved in producing X-linked colour deficiency but as yet dichromacy or anomalous trichromacy cannot be predicted from molecular patterns.

Recombinant DNA techniques have shown that photopigment genes are more numerous and more variable than expected. Many normal trichromats have more than one gene specifying long and medium-wave photopigments and although most dichromats lack the relevant photopigment gene, some individuals posses either a gene fragment or a fusion gene. These findings suggest that some genes must be suppressed. Evidence for gene suppression is obtained from studies of New World primates.

Primate colour vision

Old World primates and man have the same type of trichromatic colour vision but the colour vision of New World primates is much more variable. Many species have defective colour vision in human terms and there is large within-species variability. For example, six different phenotypes have been identified in the squirrel monkey (*Saimiri sciureus*). There are three types of trichromat and three types of dichromat. All males and some females are dichromats, combining a short-wave photopigment with one of three possible medium-wave pigments. Other females are trichromatic, combining the short-wave pigment with any two of the medium-wave pigments. It appears that three different medium-wave photopigments are potentially available at the same gene locus and that gene suppression takes place. It is supposed that male monkeys, since they have only one X chromosome, draw one pigment from the set. Homozygous females are dichromatic and heterozygous females are trichromatic. However, trichromacy can only result if both genes are expressed in the form of a mosaic and if the visual pathway is able to transmit signals from three types of cone cells as well as the usual two.

Squirrel monkeys are fructiferous animals and it may be by virtue of superior colour vision that females lead the family group in the search for ripe fruit.

References

Dalton, J. (1798). Extraordinary facts relating to the vision of colours. *Memoranda of the Literary and Philosophical Society of Manchester*, **5**, 28.

Mann, I. and Turner, C. (1956). Colour vision in native races in Australia. *American Journal of Ophthalmology*, **41**, 797–800.

Nathans, J., Piantanida, T., Eddy, R.L., and Shows, T.B. (1986). Molecular genetics of inherited variation in human color vision. *Science*, **232**, 203–32.

Scott, J. (1778). An account of a remarkable imperfection of sight *Philosophical Transactions of the Royal Society, London*, 68, 611–14.

Vanderdonck, R. and Verriest, G. (1960). Femme protanomale et heterozyote mixte avant deux files deuteranope, un fils protanomale et deux fils normaux. *Biotypologie*, **21**, 110–20.

Went, L.N. and Pronk, N. (1985). The genetics of tritan disturbance. *Human Genetics*, **69**, 255–62.
Wilson, G. (1853). Railroad signals and colour blind drivers and signalmen. Letter to the Athenaeum, London. April 2nd.
Wilson, G. (1855). *Researches on Colour Blindness*. Sutherland and Knox. Edinburgh.

Further reading

Kalmus, H. (1965). *Diagnosis and genetics of defective colour vision*. Pergamon Press, London.

6. Clinical test design and examination procedure

Colour vision test are used clinically to identify and differentiate congenital and acquired colour deficiency and to select personnel for occupations which require good colour vision. Clinical colour vision tests are simplified versions of psychophysical methods and are based on pigment colours instead of spectral stimuli. Isochromatic colour confusions and abnormal wavelength discrimination are exploited in clinical tests. Colour naming is avoided except in vocational tests.

Individual tests are designed to perform different functions (Table 6.1). Screening tests identify people with normal or abnormal colour vision. Grading tests estimate the severity of colour deficiency. Some tests have a screening and grading function. Both screening and grading tests aim to classify colour deficiency as protan, deutan, or tritan, but tests composed of pigment colours are unable to distinguish dichromats and anomalous trichromats. Vocational tests examine practical colour matching ability or colour recognition. Test manuals do not always give adequate information about the function of a test. For example, a grading test may be described in such a way as to suggest that it is useful for screening when the design is inappropriate for his function.

Tests in current use are listed in Table 6.2. There are four design categories involving different visual tasks. Pseudoisochromatic tests require the identification of a coloured figure. These tests are simple to use and have the widest application for colour vision screening. More complex qualitative judgements are demanded in hue discrimination tests, in which colours have to be arranged in sequence, and in colour matching tests. A greater degree of examiner expertise is needed to administer these tests and the normal distribution of measurements has to be determined before they can be used effectively. Lantern tests involve colour naming.

The efficiencies of most current screening tests have been established in clinical trials. Clinical trials compare the results obtained by normal trichromats and those of colour deficient observers. Colour deficient

Table 6.1 The function of different types of colour vision tests

Screening tests	Classification and grading tests	Diagnostic tests	Vocational tests
1. Anomaloscopes	1. Anomaloscopes	1. Anomaloscopes and psychophysical tests employing special stimuli	–
2. Pseudoisochromatic plates	2. Some pseudoisochromatic plates	–	
–	3. Hue discrimination (or arrangement) tests	–	1. Hue discrimination (or arrangement) tests
–	–		2. Lanterns

Table 6.2 The principal colour vision tests in each design category

Anomalocopes	Pseudoisochromatic plates	Hue discrimination (or arrangement) tests	Lanterns	Others
Nagel	Ishihara	Farnsworth D15	(a) Paired lights Holmes–Wright	TNO test City University test
Neitz Besançon	American Optical Co. (Hardy, Rand, and Rittler)	Desaturated D15 (Adams and Lanthony)	Falant	
Pickford–Nicolson	Tokyo Medical College	28 Hue	(b) Single lights Edridge–Green	Sahlgren's test
	Dvorine	40 Hue	Giles–Archer	Oscar test
	F2 Plate		Beyne	
	Velhagen	Farnsworth–Munsell 100-hue	Colour-threshold tester	
	Ohkuma			
	American Optical Co. (13375)			
	Bostrum–Kugelberg			
	Standard pseudoisochromatic plates 1st and 2nd editions			
	Lanthony tritan album			

people acting as observers in trials are classified from psychophysical measurements or from a Rayleigh match. False positive and false negative rates are determined for each test or for each part of a test. Low values indicate an efficient test. 'Sensitivity' and 'specificity' can also be calculated. 'Sensitivity' is the percentage, or proportion, of colour deficient people correctly identified and 'specificity' is the percentage of colour normals correctly identified. High values indicate an efficient test. Results can be analysed to show test/retest reliability and coefficients of agreement between tests. However, these data should be accepted with caution since differences in administration and interpretation can produce large variations in the statistical measure of agreement and tests based on different design criteria should not be expected to show close agreement. Tests based on visual performance show a learning effect which influences test/retest reliability.

Screening efficiency is sometimes derived from the ability of a test to identify the correct percentage of colour deficient males in an unselected population. A large number of people have to be examined to obtain the required accuracy of measurement. A smaller number of observers may be sufficient if the efficiency of a new test is compared with that of an established one.

Congenital tritan defects are rare and a number of tests are designed for red–green colour deficiency only. The efficiency of new tritan tests is usually determined from the examination of patients with acquired colour deficiency. Tests are not usually designed specifically for acquired defects, diagnosis of acquired colour deficiency is made from an accummulation of test results and from the medical history.

The design of pseudoisochromatic plates

Pseudoisochromatic plates are examples of colour camouflage. The individual elements of the designs are spots or patches of colour. These are selected and positioned in such a way that a figure can be seen by people with normal colour vision but cannot be seen by people with defective colour vision. The dot matrix breaks up the outline of the figure and conceals it's shape. The colours contained in the figure and in the background are isochromatic pairs and all the coloured spots appear 'falsely of the same colour'. Screening plates for detecting red–green defects are usually composed of orange, yellow, and yellow–green spots, because isochromatic zones for protans and deutans are similar in this quadrant of the chromaticity diagram. 'Vanishing' designs, in which a figure is concealed, are the simplest to construct and all pseudoisochromatic tests contain designs of this type. Some tests, particularly the Ishihara plates, contain more complex patterns. By careful placement of the colours, a figure can be distinguished by people with defective colour vision which cannot be seen by normal trichromats. In this case the figure is camouflaged for normal observers. This is achieved by selecting colours for the background matrix of dots which look the same to colour deficient people but which break up the outline of the figure in normal colour vision. This principle may be used in separate pseudoisochromatic patterns or can be superimposed on a vanishing design. In the latter, an apparent transformation of the perceived figure takes place: normal trichromats see one figure and colour deficient people see a different figure in the same design. Transformation designs give positive evidence of colour deficiency as opposed to negative evidence obtained from vanishing designs. This is a distinct advantage when examiners are unsure whether failure to see a vanishing design is due to colour deficiency or poor comprehension of the visual task.

Pseudoisochromatic plates have to be carefully constructed to be effective. The selected colours should have chromaticities within the appropriate dichromats isochromatic zone. This can be checked by measuring the spectral reflectance of the individual colours and by calculating the CIE chromaticity coordinates. Alternatively, printed colours can be matched with Munsell samples and the CIE coordinates obtained from published data (Wyszecki and Stiles 1967). The latter method has the advantage that the colours can be matched under the same lighting conditions as those employed for the test and any change in appearance, due to simultaneous colour contrast, is automatically taken into account. Pigment tests are designed to be illuminated by CIE source C. Luminance contrast and colour differences between pairs of isochromatic colours are also calculated. The size of the colour difference step determines the level of difficulty of the discrimination task. If this is inappropriate, either false positive or false negative results will be obtained. Small colour difference must be employed in screening plates. However if the colour difference step in vanishing designs is too small, some normal trichromats cannot see the figure and are incorrectly diagnosed as colour deficient. Conversely, if the colour difference step is too large, some colour deficient observers identify the figure and are incorrectly classified as normal trichromats. Chromaticity coordinates and colour difference steps have been calculated by Lakowski and his co-workers for most pseudoisochromatic tests in general use. In the earlier reports, colour differences were calculated using the Judd Hunter formula (1942) which expressed colour differences in terms of NBS (National Bureau of Standards) units. More recently measurements have been made with the CIELAB equation (1978) which is derived from uniform colour space.

Pseudoisochromatic designs intended to grade the degree of colour deficiency employ several colour difference steps. Ranking the degree of colour deficiency is achieved, in effect, by measuring colour confusions represented by isochromatic lines of different length (Fig. 6.1). For red–green defects, the values of ΔE measured by the CIELAB formula are about 15–22 for screening plates, 30–40 for detecting moderate defects, and 50–60 for severe defects. Classification designs to distinguish protan and deutan defects are based on neutral colours (colours confused with grey) and have a large value of ΔE. The background dot matrix is composed of greys which vary in lightness (Munsell value), and the colours chosen for the figures are the average neutral colours for each type of defect. Red–purples are preferred to greens because there is a larger difference between the average protan and deutan neutral colours. Even so, if a high density of macular pigment is present, the colour confusions of both protanopes and deuteranopes include the average neutral colours for both types of deficiency and neither figure can be seen. These individuals may be incorrectly diagnosed as monochromats if no other tests are used.

Normal variations in threshold blue perception make it difficult to design efficient pseudoisochromatic tests for tritan defects. If small colour

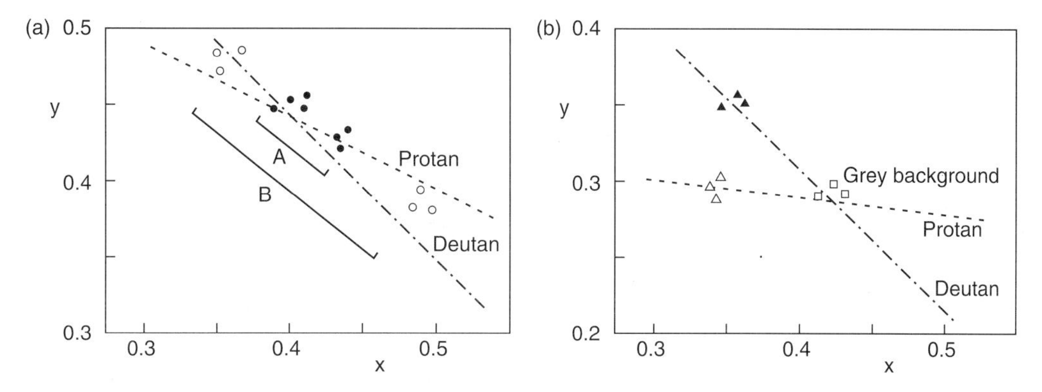

Fig. 6.1 Colours selected for pseudoisochromatic tests compared with isochromatic zones for protan and deutan defects. Colours selected for the background matrix of dots and for 'vanishing' figures have coordinates within isochromatic zones for the particular type of colour deficiency for which the design is intended. (a) Small colour differences are employed in vanishing designs for the detection of red–green colour deficiency (A) and large colour differences are used to distinguish severe defects (B). (b) Neutral colours (colours confused with grey) are used to differentiate protan and deutan defects. The background consists of grey dots (□) and the diagnostic colours are blue–green for protans (Δ) and green for deutans (▲).

differences are employed in screening tests, a large number of false positive results are obtained. The number of false positives increases with age, due to physiological changes in the optic media. Designs based on large colour differences do not identify people with slight tritanomalous trichromatism but produce a smaller number of false positive results. For tritan defects the corresponding values of ΔE are about 30–40 for screening tests, 60–70 for moderate defects, and 90–100 for severe defects. Neutral colours in the violet part of the spectrum are usually employed for tritan classification plates because it is difficult to maintain the correct luminance contrast relationship between greys and yellows.

Pairs of colours within isochromatic zones only appear identical if no luminance contrast exists, and some pseudoisochromatic designs give false negatives because insufficient attention has been given to this factor. Relative luminous efficiency is different in protan and deutan defects and designs intended to screen for both types of deficiency contain several chroma and lightness differences to give a range of luminance contrast values. Printed designs must be carefully registered. If the printing is slightly out of register superimposed dots and unintentional gaps make the outline of the figure easier to detect. Random and regular dot matrices have both been used in pseudoisochromatic designs. The former allow more flexibility in the design of the contained figure but present a more complex printing exercise. The size of the spots, and the size and stroke width of the figure, vary in different tests and different levels of acuity are necessary to complete the visual task. Tests which have small spots and narrow figures are ineffective if visual acuity is slightly reduced while others are more robust (Table 6.3). Visual acuity has to be carefully considered when selecting suitable tests for patients with acquired colour deficiency.

Table 6.3 Level of visual acuity required for the interpretation of pseudoisochromatic plate tests

Text	Minimum visual acuity required for interpretation		
Bostrum–Kugelberg	6/9	0.66	20/30
Tokyo Medical College	6/12	0.50	20/40
Ishihara	6/18	0.33	20/70
American Optical Company (Hardy, Rand, and Rittler)	6/60	0.10	20/200

Pseudoisochromatic tests are durable and easy to use. Tests last for a number of years if handled carefully. However any book test becomes dilapidated with repeated use and has to be replaced. Replacement is also necessary if the plates get dirty. Fading of the designs is not a problem if the book is closed after use and stored away from direct sunlight or in the box provided.

Pseudoisochromatic plates are sometimes described as PIC tests.

The administration of pseudoisochromatic plates

Pseudoisochromatic plates are designed to be viewed at two-thirds of a metre or at 'arm's length'. The spots of colour have the desired angular subtends at this distance. The examiner turns the pages of the book so that the viewing time is carefully controlled. The visual task is simple and little introduction is necessary. The examiner may say 'Tell me the numbers you can read as I turn the pages. Sometimes you may not see a number and then I will turn to the next design.' Some people become concerned if they cannot see a figure on every plate so it is a good idea to mention this possibility beforehand. An introductory plate is often included to demonstrate the visual task. A correct response to the introductory plate shows that the person has sufficient visual acuity to complete the test and is not malingering. The observer looks directly at the design and must name the figure immediately. About four seconds is the maximum time allowed for each plate. If the figure is not seen in the allotted time, the examiner turns to the next design. Undue hesitation can be a sign of slight colour deficiency and if the examiner is uncertain how to interpret slow responses, confirmation should be sought from other tests. It is useful to record the responses to each design. This helps to evaluate individual viewing conditions and serves to identify designs which produce slight interpretive mistakes. Isolated interpretive mistakes, such as '6' being described as '8' on a vanishing design, should not be considered as a failure of the plate or as an indication of colour deficiency. If time is limited, a test can be abbreviated by showing a selection of the most efficient plates but reliance on a single plate is not desirable. It is sometimes necessary to change the order of the plates to prevent the correct response sequence being learned. People who wish to join a profession or recreational activity which requires normal colour vision may attempt to learn the correct results. On the other hand, military conscripts and children with psychological problems may try to simulate colour deficiency. In these rare cases the examiner is alerted by the demeanour of the observer and by the inconsistency of the results.

Some pseudoisochromatic tests have designs containing 'pathways' for the examination of nonverbal subjects and young children. The observer traces the pathway using a clean paintbrush or cotton bud. Alternatively, cutout or printed replicas can be provided for figure designs and the person selects the matching figure or turns the replica to match the one seen. A longer viewing time is needed for these procedures and this slightly reduces the screening efficiency of the test. A prolonged viewing time allows the person to scan the design and to alter the angle of viewing. This enhances the visibility of chromatic contours and aids detection of the concealed figure.

Computer controlled displays have been used to examine congenital and acquired colour deficiency. Pseudoisochromatic patterns, containing isochromatic colour pairs or neutral colours, are displayed on a raster screen. The computer is programmed to change the saturation and luminance contrast in steps and the observer presses a button when a pattern is seen. The colour range is limited by the screen phosphors and it is not possible to display fully saturated colours. The technique is very effective in demonstrating isochromatic zones but cannot be used to distinguish dichromats and anomalous trichromats.

The design of hue discrimination or arrangement tests

Hue discrimination tests were originally designed for vocational guidance. They are grading tests which identify moderate and severe colour deficiency and classify protan, deutan, and tritan defects. Qualitative judgements are needed to arrange the colours and the tests are unsuitable for young children or educationally disadvantaged groups. The use of Munsell colours enables the chromaticity coordinates of each colour to be identified and colour differences to be calculated. Hue discrimination tests are particularly useful for the examination of acquired colour deficiency because nonspecific defects can be identified and changes with time recorded. Colour arrangement tests require manual dexterity and must be placed on a table. The person is encouraged to hold the cap and not to touch the colour surface, but many people cannot avoid doing so and the colours eventually become dirty and have to be replaced.

The most widely used hue discrimination tests: the dichotomous (or D15) test and the Farnsworth–Munsell 100 hue test (F–M 100 hue test), were designed by Dean Farnsworth in 1943. Both tests contain colours selected from a complete hue circle (Fig. 6.2). The individual colours are held in a circular cap subtending 1.5° at a test distance of 50 cm. The size

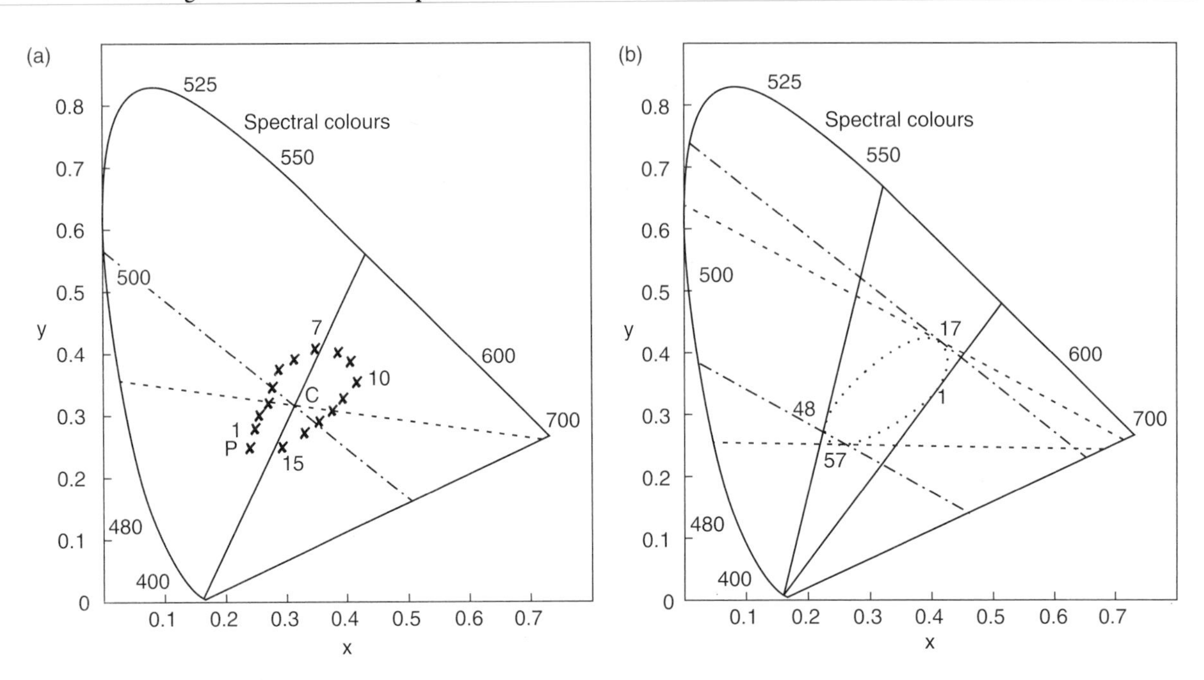

Fig. 6.2 Chromaticity coordinates of the samples used in Farnsworth–Munsell tests. (a) The Farnsworth D15 test. The 16 Munsell samples of the test represent a circle of equal value and chroma. Colours from opposite sides of the circle which have coordinates within isochromatic zones have similar colour appearance to colour deficient observers and may be mingled in the arrangement of colours. (b) The Farnsworth–Munsell 100-hue test. Isochromatic zones for protan, deutan, and tritan defects form tangents to the chromaticity locus of the 85 Munsell samples of the test. Colour confusions, misarrangement of the colours, occur in two regions which correspond with these zones and are almost diametrically opposite in the polar diagram used to score the test. A typical 'axis of confusion' results in each type of colour deficiency.

of the individual colour caps ensures that observations are made with the central rod-free retinal area. All the colours of the D15 test are presented together in a single box and isochromatic colour confusions are demonstrated when colours from opposite sides of the hue circle are placed next to each other in the arrangement. Several D15 panel tests, having different Munsell value and chroma, are in current use. The City University test is derived from the D15 but displays a small selection of the colours on pages in a book. In these tests the colour difference step varies according to the number of hues and the saturation level. In the standard test, the value of ΔE varies between 5 and 13 for adjacent colours in the sequence and isochromatic confusions have a ΔE of about 40.

The purpose of the F–M 100 hue test is to measure hue discrimination ability at constant value and chroma. A special series of Munsell samples is used which have smaller hue differences than those available in the standard Munsell system. This means that individual hues cannot be identified using Munsell notation and lost or soiled colours can only be replaced by reference to the F–M 100 hue number. The value of ΔE between adjacent hues is less than 3. The small colour difference steps make the test difficult for normal trichromats to complete without making some mistakes. The colour differences between each of the 85 colours are not quite the same. Smaller differences occur in the blue–green quadrant of the hue circle and it is more likely that mistakes are made with these colours. These variations are reproduced as differences in the spacing of the radial lines of the polar diagram used to plot the test results. Most computer drawn diagrams do not retain these differences and the appearance of typical plots, showing different types of colour deficiency, is slightly changed. Isochromatic colour confusions cannot be demonstrated with the F–M 100 hue test because the colours are divided into four boxes and colours from opposite sides of the hue circle are not presented at the same time. In protan, deutan, and tritan defects, poor discrimination of adjacent hues occurs in two nearly opposite regions in the hue circle. These regions are located where isochromatic zones are tangential to the locus of hues. The two clusters of errors give rise to an apparent 'axis of confusion' representing different types of colour deficiency. The prominence of the axis of confusion coupled with the error score gives an estimate of the severity of the discrimination deficit.

Two shortened tests have been developed from the F–M 100 hue test. The 28 hue test is constructed by taking approximately every third colour from the F–M 100 hue sequence, and the 40 hue test takes approximately every second colour. All the colours are presented together and both isochromatic colour confusions and poor discrimination of adjacent hues can be made. The results are often difficult to interpret and neither test has been widely used.

The administration of hue discrimination tests

To begin the test, all the colour caps, except the reference colours, are removed from the box, mixed up, and placed in a group in front of the observer. The observer then puts the colours back into the box in what he perceives to be a natural order, starting either from a single reference colour or arranging the colours between two reference caps. For example 'Please put the colours back into the box so that they form a natural colour order or appear to change gradually in steps.' A step-by-step procedure may be easier to understand. 'Find the colour which looks most like the colour in the box and put it next to it. Then find the colour which looks most like the last one placed in the box and so on. When the colours are all in the box, look along the order again and see if any changes need to be made so that the colours are arranged gradually in steps.' Although the examiner need not be present during the examination, more meaningful results are obtained from personal testing. If the examiner watches the colours being arranged initial mistakes can be observed even if these are subsequently corrected. Unlimited viewing time is usually allowed but the examiner notes how carefully the test is completed. The four boxes of the F–M 100 hue test are always presented in rank order beginning with box 1. This enables the examiner to appreciate loss of concentration or improvement in performance as the person becomes more familiar with the visual task. After the final review of each box, the number order on the back of the caps is recorded, the error score

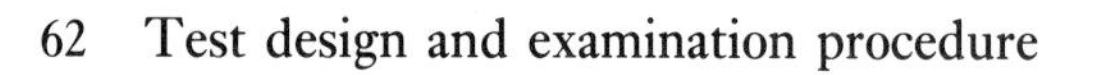

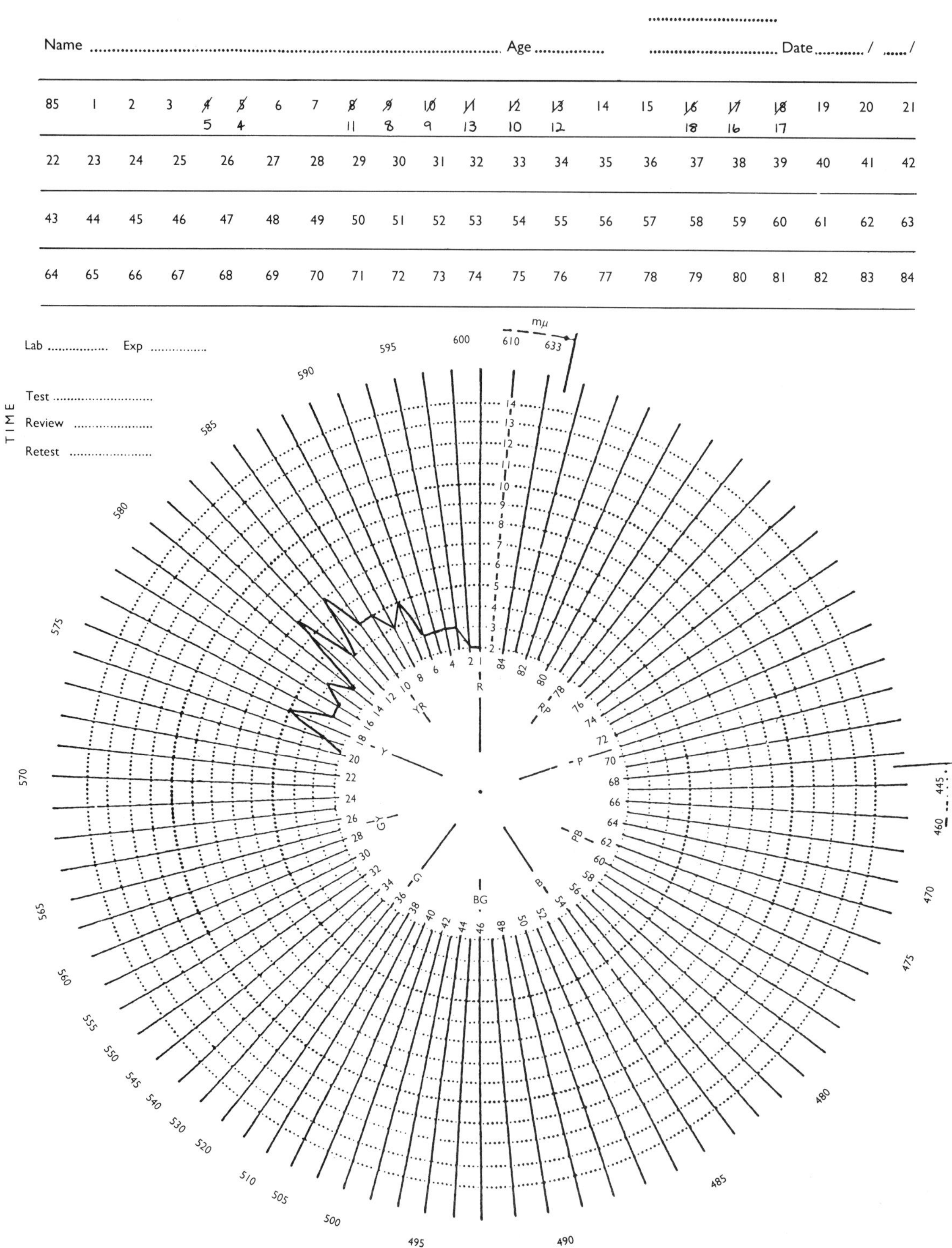

Name .... Age .... Date ..../ ....../

85	1	2	3	~~4~~	~~5~~	6	7	~~8~~	~~9~~	~~10~~	~~11~~	~~12~~	~~13~~	14	15	~~16~~	~~17~~	~~18~~	19	20	21
				5	4			11	8	9	13	10	12			18	16	17			
22	23	24	25	26	27	28	29	30	31	32	33	34	35	36	37	38	39	40	41	42
43	44	45	46	47	48	49	50	51	52	53	54	55	56	57	58	59	60	61	62	63
64	65	66	67	68	69	70	71	72	73	74	75	76	77	78	79	80	81	82	83	84

Lab .... Exp ....

TIME
Test ....
Review ....
Retest ....

calculated, and the results plotted graphically. Colour vision clinics and laboratories use computer programs to assist this process. Departures from the normal cap sequence are entered manually and the computer calculates the error scores and displays the resulting polar diagram on the screen. A wide range of programs are available including one for personal computers using Windows software. Printed copies of the results can be made if desired. Results for groups of observers can be stored on computer and retrieved later for statistical analysis. Systems which interface the F–M 100 hue test directly with a computer have been developed. These have the advantage that the colour order is entered into the computer program automatically rather than manually.

The F–M 100 hue results are plotted on a polar diagram representing the 85 hues of the test. The score for each cap is calculated as the sum of the numerical differences of the adjacent caps in the arrangement, and is plotted on the radial line of the polar diagram representing that colour (Fig. 6.3). The baseline of the diagram represents the correct score of 2 and this value is deducted from the individual score to obtain an error score. The total error score for the test is the sum of the individual errors. The total error score therefore represents the area under the curve displayed on the polar diagram. Classification of the type of colour deficiency is made from the appearance and position of the axis of confusion. Averaging methods of computation have been developed to assist visual interpretation of the axis of confusion. These methods are particularly useful in the study of acquired colour deficiency when the error score is large and the axis of confusion is difficult to interpret.

An alternative method for plotting the F–M 100 hue results has been suggested by Kinnear (1970). This method treats the test as one of colour

---

Fig. 6.3 Scoring and plotting the Farnsworth–Munsell 100 hue test. The score for each colour (number cap) is calculated as the numerical sum of the differences between the preceding and following caps. The baseline of the polar diagram represents a score of 2. The score for each cap colour is plotted on the radial line of the polar diagram which represents that cap number. The error score is the score with 2 subtracted and is shown by the number of scale divisions above the baseline. The total error score fot the test is obtained by summing the individual error scores.

An individual test result is recorded by striking out relevant numbers in the sequence and substituting the actual colour placed by the observer. For example, for the first box of the test:

Correct sequence 7 8 9 10 11 12 13 14

Actual sequence 7 11 8 9 13 10 12 14

Examples of scoring in this actual sequence:

Cap 8, score 4 (plotted), error score $4 - 2 = 2$.

Cap 13, score 7 (plotted), error score $7 - 2 = 5$.

The total error score is the sum of the individual errors (or the number of scale divisions above the baseline). In the example shown, the total error score is 36.

Table 6.4 Limits of the normal error score obtained with the Farnsworth–Munsell 100 hue test viewed binocularly and illuminated by CIE source C giving 200 lx

Age range in years	Maximum normal score score at the 95% confidence limit	Square root of the error score Mean	Standard deviation
10–14	193	9.13	1.85
15–19	122	6.63	1.91
20–29	107	5.69	2.07
30–39	133	6.71	2.90
40–49	188	8.23	2.44
50–59	234	8.68	2.64
60–69	268	9.57	2.44
70–80	317	11.46	2.01

Derived from Verriest *et al.* 1982.

sequence placement and the individual errors are plotted in the polar diagram in serial order as they occur in the arrangement. This does not introduce any significant change in the axis of confusion and the total error score is unaffected.

Normal trichromats often make several randomly placed errors on the F–M 100 hue test. Normal hue discrimination ability is optimum during the second decade and declines slightly with age. Deterioration of this ability, especially after the age of 50 years, can be shown by a increase in the upper limit of the normal error score. Errors cluster in the blue–green part of the hue circle because the colour differences between the caps are slightly smaller in this quadrant of the test and because threshold blue perception declines with age. Age dependent critical values have been determined using an illuminance of 200 lux (lx) (Table 6.4). This luminance level is lower than the recommended value and the larger error scores which result show the effect of age particularly well. Error scores decrease slightly as the illumination level is increased and the figures quoted in Table 6.4 are higher than those achieved in practice. Error scores are also larger if no final review and correction of the colour arrangement is permitted. This may have contributed to the high error scores in Verriest's data. Subtle differences in the administration of the F–M 100 hue test, even the attitude of the examiner, may affect the error score and comparisons of data obtained by different examiners should be treated with caution.

The results of the D15 test are simple to transfer to the record sheet and no additional computation is necessary. The examiner draws a line joining the cap numbers in the order in which the colours have been placed. The record sheet provides an aid to interpretation by illustrating typical isochromatic confusions in protan, deutan, and tritan defects. Results obtained with the 28 hue test and the 40 hue test are recorded in the same way.

Numerical scoring methods, based on colour difference steps, have been developed for the D15 test to assist the analysis of acquired colour deficiency.

Illumination

The appearance of pigment colours changes with the spectral content of the illuminant. Colours which are isochromatic when illuminated with source C may be altered to lie outside isochromatic zones if a different illuminant is used. For example, tungsten illumination alters the appearance of the Ishihara plates and enables some slight deuteranomalous trichromats to obtain the correct result. A CIE standard source C is required for optimum results with all clinical colour vision tests based on pigment colours. The MacBeth easel lamp has been specially designed for plate tests and is universally employed in clinical trials to determine their reliability. The MacBeth lamp consists of a 100 W tungsten bulb fitted with a hemispherical blue filter which converts the spectral distribution to source C. This assembly provides between 300–400 lx (approximately 100 candelas (cd) m^{-2}) on the surface of a plate test positioned in the tray below the light source. Use of the tray ensures that the illumination is incident at the desired 45° to the plate surface. The blue filter may vary in thickness and the exact illuminance should be measured for each lamp. Unfortunately manufacture of the MacBeth easel lamp has now been discontinued.

The MacBeth lamp or normal daylight (facing north in the northern hemisphere and south below the equator) are preferred illuminants for colour vision tests. Colour corrected fluorescent tubes and industrial colour matching cabinets are satisfactory alternatives. Fluorescent light sources should have a colour temperature of about 6500 K, a balanced spectral distribution and a colour rendering index over 90. The Thorn Daylight fluorescent tube fulfils these requirements. Industrial colour matching cabinets, which contain several light sources which can be individually adjusted in intensity to give source C, are satisfactory for colour vision examinations. Test manuals do not normally specify the required illuminance for optimum performance, but few examiners use less than 250 lx or more than 650 lx. Colour constancy occurs within this range. It is important to maintain a constant level of illuminance if a series of tests, containing saturated and desaturated colours, are being used to grade the degree of colour deficiency. Results obtained in different locations can be compared only if the same light source and level of illuminance is used.

The design of anomaloscopes

Spectral anomaloscopes have been developed specifically for the diagnosis of different types of congenital colour deficiency. A spectral anomaloscope is a psychophysical test which presents one or, in some cases, two diagnostic colour matches. The need to provide only a small number of wavelengths reduces the size of the apparatus and makes it reasonably portable. Accurate calibration is required and the observation parameters are carefully specified. The two colour matches presented are the Rayleigh match, red mixed with green to produce yellow, and the Engelking–Trendelenberg match, blue mixed with green to match blue–green. The bipartite matching field has an angular subtend of approximately 2°. This ensures that foveal cones are stimulated and that the distribution of macular pigment is uniform. The characteristics of the Rayleigh match determine whether a person has normal or defective red–green colour vision; whether the defect is protan or deutan; and whether the person is a dichromat or anomalous trichromat. Several different wavelengths can be used successfully in a Rayleigh match. However, the most reliable results are obtained if the matching wavelengths are widely separated on the straight line section of the CIE spectral locus. Emission lines are chosen for ease of calibration. The sodium line at 589 nm is selected for the yellow test wavelength. Normal hue discrimination is optimal around this value and threshold sensitivity is invariant with age. In the Nagel anomaloscope the test field is 589 nm, the red mixture wavelength is 670 nm, (lithium) and the green wavelength is 546 nm (mercury). The wavelengths for a blue–green anomaloscope have to be carefully selected to minimize the effect of population variance due to macular pigment density. There is no straight line along the spectrum locus in the blue–green part of the spectrum. A mixture of blue and green wavelengths produces a desaturated colour and a desaturating wavelength must be added to the blue–green test field to obtain an exact colour match. Different amounts of the desaturating wavelength need to be added as the ratio of the mixture colours is changed. The matching procedure is therefore more complicated than for a Rayleigh match. The prototype Besançon anomalometer incorporated an Engelking–Trendelenberg match with 489 nm (blue–green), 516 nm (green), and 470 nm (blue). The wavelengths are produced by interference filters. After 1983 the blue–green match was changed to the ideal combination specified by Moreland and Kerr (1979) to compensate for different densities of macular pigment. The desaturated test field contains a fixed mixture of blue–green (480 nm) and yellow (580 nm), and the matching primaries are 436 nm (blue) and 490 nm (green). This is known as the Moreland match and has been specifically developed for the analysis of acquired colour deficiency.

Anomaloscopes, such as the Pickford–Nicolson anomaloscope, which contain broad-band filters, do not have the same screening or diagnostic capabilities as spectral instruments. The use of glass or gelatin filters always produces desaturated mixture colours and the luminance of the test field varies for different red–green ratios.

Administration of anomaloscopes

Administration of an anomaloscope test requires knowledge and skill. The normal matching range has to be established before an individual instrument can be used clinically. The test procedure begins by demonstrating the instrument controls. The observer then manipulates both the red/green ratio and the yellow luminance controls until the two halves of the test field look exactly alike. 'Alter both the control wheels until the two halves of the circle look exactly the same colour and the same brightness.' Colour names need not be used. Four separate matches are made. The first match is usually considered as a learning exercise and only the last three matches are recorded. About ten seconds is allowed for each match and in between matches the observer looks away from the instrument into a dimly lit room. This reduces chromatic adaptation and minimizes after images. The observers initial colour matches are used as a guide to the second, and most important, part of the examination procedure which is to determine the limits of the matching range. The examiner sets the red/green mixture ratio in steps and the observer ascertains whether an exact match can be obtained using the yellow luminance control only. 'I have set one half of the field of view. See if you can obtain an exact match by altering this wheel only. Please answer either "Yes", I can match the two halves of the circle exactly or "No", I cannot.' The scale readings at the limits of the matching range are recorded, compared with the normal limits, and a diagnosis made. Each eye is examined separately.

Different methods for presenting Rayleigh colour matches have been developed including the use of light emitting diodes and displays on a raster screen. In computer driven displays a ranked series of red–green ratios is presented and the person responds by pressing a button to indicate whether the two halves of the matching field are identical or not. Each presentation is limited to a few seconds. In some systems, the observer makes a qualitative judgement and presses alternative buttons to indicate whether the mixture field is too red or too green. The computer offers red/green ratios in a 'staircase' procedure by offering smaller and smaller increments until the limits of the matching range have been determined.

Although, colour naming is usually avoided, naming can be of assistance in confirming the results obtained with young children or educationally disadvantaged groups. Normal matches are contained within a small range of red/green ratios. If the normal mean match is presented, people with normal colour vision should say that the two halves of the matching field are either the same or very nearly so. Protanomalous trichromats should report that the mixture field looks 'too green' and deuteranomalous observers that it looks 'too red'. The examiner then makes appropriate adjustments, based on knowledge of the normal matching range, and asks for repeated colour naming until the limit of the matching range has been determined. Dichromats may accept the normal match and it is necessary to present several red/green ratios, with appropriate luminance values, in order to make the correct diagnosis.

Some anomaloscopes incorporate white illuminated screens for the person to view between matches. This is intended to provide neutral adaptation but may actually prevent recovery from chromatic adaptation and make the person's colour matches unstable. Viewing a dark area is preferred.

The design of colour vision lanterns

Colour vision lanterns are vocational tests which employ colour naming. Lanterns are designed to test the recognition of coloured light signals used in maritime, military, aviation, and transport services. Lanterns simulate the visual task required in these occupations. The Edridge–Green lantern and the Farnsworth (Falant) lantern are exceptions. The Edridge–Green lantern contains a large number of colours which are not used for signals and the colours incorporated in the Falant are isochromatic colours not signal colours.

There are two types of lanterns, those which show colours in pairs and those which show single colours. Paired colours, representing port and starboard navigation lights, are usually shown separated vertically rather than horizontally. Some lanterns have three apertures in a triangular pattern to represent port, starboard, and mast-head lights, but these are not often used. The colours are reproduced by the same filters as those used for the actual signals.

A lantern test is either given in normal daylight (photopic) viewing conditions, or after a period of dark adaptation to represent night-time (scotopic) observation. Both viewing conditions are used in a comprehensive lantern test.

Some lanterns have different sized apertures to represent signal lights at different distances, and some of the older lanterns have frosted and ribbed glassed to simulate fog and rain. All lanterns contain red, white, and green filters, and most lanterns contain two different reds and two different greens, both of which are within the internationally agreed specifications for signals. Lanterns which show pairs of colours have a programmed presentation and detailed operating instructions are supplied with the instrument. This ensures that the examination is always conducted in the same way.

If operated carefully a lantern test can be an efficient screening test for red–green colour deficiency. However, pass/fail criteria, based on the type of error or the number of errors, are often a poor guide to the type and severity of colour deficiency and other clinical tests should be used if this information is required. If a lantern test is carried out with small low intensity lights after dark adaptation, about 30 per cent of normal trichromats have difficulty in differentiating green and white. This result is due to night myopia.

Lanterns which show single colours include yellow as well as red, green, and white. These lanterns are primarily designed for railway workers. Railway signals are usually shown singly and a yellow light designates 'proceed with caution'. Coincidentally, the inclusion of yellow leads to better protan/deutan classification since deutans tend to call yellow 'red' and protans tend to call it 'green'. A dark red is often included to represent a distant red signal at night. This is very effective in demonstrating 'shortening of the red end of the spectrum' as protans report that no light is being shown. The order of the colours is not usually fixed and expertise is needed to administer the test accurately and consistently.

Administration of lanterns

Lanterns are usually viewed at a distance of 6 m or 20 ft. An ophthalmic consulting room is ideal as the examiner can stand next to the observer while operating the lantern, and the colours are viewed through the mirror positioned 3 m away. It is important to explain that only certain colour names are allowed before the examination begins. These are either red, green, and white, or red, green, white, and yellow depending on the lantern. Several colours are shown and named correctly by the examiner to demonstrate the visual task before the test begins. The person must name the colour or colours immediately they are shown. If colours are shown in pairs, the top one is named first. Each colour or colour pair is displayed for about 5 s. The examiner quickly records the responses on a prepared record sheet (using initials for speed), and then presents the next colour in the sequence. At least 20 colour pairs or single colours are shown in both photopic and scotopic viewing conditions. Repeating each colour several times is essential to demonstrate false colour naming in colour deficient people.

Vocational tests

Vocational tests reproduce the visual task of a particular job and aim to determine whether a person can complete it satisfactorily. Most vocational tests have a grading function. Colour deficient people who fail a screening test but pass a vocational test are considered to be capable of doing the job satisfactorily. Vocational tests may also serve to identify people with normal colour vision and superior colour aptitude.

A 'resistor wire test' is sometimes given to recruits in the electrical and electronics industry. There is no formal test of this type and each test has to be devised by the examiner. An examination may consist of asking a recruit to name the coloured bands on a number or resistors, or to match the wires in a multicore cable. This informal procedure is unlikely to be effective either as a screening or grading test. A degree of standardization can be attempted by mounting the resistors on a board and recording the colour names given. A comprehensive test containing a range of samples would take about 15 min and has no advantage over standard clinical tests except relevance to the occupation.

Two tests designed to show superior colour aptitude were designed in the 1930s and were used extensively in the US. These were the Intersociety Colour Council colour aptitude test and the Burnham, Clark, Newhall hue memory test. The colour aptitude test involved both hue and saturation discrimination. Four Munsell hues (blue, red, green, and yellow) were presented in 10 saturation steps and the recruit was asked to match each colour by selecting the identical sample from the complete set of colours. An error score was recorded and people were classified as either excellent, good, satisfactory, or doubtful. The hue memory test consisted of 20 Munsell hues and 43 comparison hues mounted on two concentric discs. Each of the 20 test hues was shown for 5 s. The recruit then attempted to select the matching hue, from memory, by rotating the comparison disc and viewing each colour individually. An error score was calculated and the person's colour aptitude classified as either superior, normal, or poor. Correlation between the classification categories and the needs of a particular occupation was decided empirically.

The modern approach to vocational testing is to offer advice on the results of a clinical test battery and on the error score obtained on the F–M 100 hue test if superior colour aptitude is required.

References

Farnsworth, D. (1943). The Farnsworth–Munsell 100 hue and dichotomous test for color vision. *Journal of the Optical Society of America* **33**, 6568–78.

Hunter, R.S. (1942). A photoelectric tristimulus colorimeter with three filters. *Journal of the Optical Society of America*, **32**, 509.

Kinnear, P.R. (1970). Proposals for scoring and assessing the 100-hue test. *Vision Research*, **10**, 423–33.

Moreland, J.D. and Kerr, J. (1979). Optimisation of a Rayleigh type equation for the detection of tritanomaly. *Vision Research*, **19**, 1369–75.

Verriest, G. Laethem, J. van, and Uvijls, A. (1982). A new assessment of the normal ranges of the Farnsworth–Munsell 100 hue test scores. *American Journal of Ophthalmology*, **93**, 635–42.

Further reading

Lakowski, R. (1969). Theory and practice of colour vision testing. *British Journal of Industrial Medicine*, **26**, 173–89, 265–88.

National Research Council (Working Group 41) (1981). *Procedures for testing color vision*. National Academic Press, Washington DC.

Wyszecki, G. and Stiles, W.S. (1967) *Color science*. John Wiley, New York.

7. Tests for defective colour vision

The first method for examining colour vision consisted of colour naming. Huddart questioned Harris about the appearance of everyday objects and disagreement about colour names first alerted John Dalton to his own colour deficiency. Dalton's observations of spectral colours led him to discover an achromatic band in the middle of the spectrum and reduced sensitivity to long-wavelength light compared with other people. Dalton assumed that all colour deficient people were alike and used these two phenomena as the basis for laboratory colour vision tests. Dalton's apparatus could not be moved so he made a simple test, consisting of coloured ribbons, to examine the surviving members of the Harris family. Dalton's coloured ribbon test is the first recorded clinical text for colour deficiency. The idea of arranging colours in sequence or sorting them into groups was introduced by August Seebeck in 1837.

Seebeck's test consisted of 300 coloured papers and must have taken a long time to complete. Even so, this was only part of the examination procedure and he measured regions of maximum and minimum sensitivity in the spectrum as well. The results of this comprehensive investigation led Seebeck to conclude that there were two distinct classes of red–green colour deficiency which could not be distinguished by colour naming. He found a sequence of severity in both classes thus anticipating the work of Rayleigh fifty years later.

In spite of the superiority of Seebeck's examination methods, colour naming continued to be used for screening large groups of people and this was the method used by Wilson in 1853. Wilson's survey confirmed that the incidence of colour deficiency was the same in all social classes and could not be attributed to lack of education as suggested by some of his contemporaries. Wilson speculated that accidents could result from mistakes made by people in key occupations and specifically recommended that colour deficient men should not be employed as signalmen and engine drivers. The great expansion of railway systems had just begun and Wilson's remarks proved a timely warning. Some measures seem to have been introduced by French railway companies as early as 1858 and colour vision testing became firmly established after 1864. Results obtained by railway doctors confirmed the high incidence of colour deficiency, and the French physician A. Favre urged that colour vision testing be undertaken in schools and that colour vision standards be introduced in the army and navy. Concern about the safe running of the railways appeared to be fully justified when a collision occurred in the early morning of 15 November 1875 near the town of Lagerlunda in Sweden. Although the accident could be attributed to failure to stop at a red light, all the people directly involved were killed and there was no proof of the exact cause. Nevertheless the Swedish railway authorities decided that all employees must be able to pass a colour vision test and they commissioned one to be designed. The test chosen was the Holmgren wool test. Other national railway authorities also adopted the Holmgren test, but sorting coloured wools did not seem particularly relevant to recognizing signal lights and the first colour vision lantern was introduced by Edridge–Green in 1891.

At the same time a different method of colour vision testing was being developed in Germany and the first painted pseudoisochromatic charts were introduced by J. Stilling of Strasbourg in 1876. The colours used in Stilling's vanishing pattern designs were derived from colour mixtures made by two colour deficient observers. Stilling supported the Hering theory of colour vision and his test provided for four types of colour deficiency. The Stilling plates were later redesigned by Hertel and further derivatives by Velhagen and Broschmann are in use today.

The introduction of these different methods of examination produced a number of other individually designed matching and sorting tests. Some of these had intriguing names such as the 'Philip dominant colour test', 'Oliver's worsted test', 'the Edridge–Green bead test', and 'Wiltberger's horizontal colour strips test'. The development of colour printing in the second decade of the twentieth century opened the way for mass production of printed pseudoisochromatic charts and other tests became less popular. The Ishihara pseudoisochromatic plates first appeared in 1917 and rapidly became the standard test for detecting red–green colour deficiency throughout the world. The test has been reprinted many times. The first fifteen editions are in rank numerical order but since 1962 each new edition is identified by the year of publication. The colour printing varies slightly in different editions but this has very little effect on test performance. Unlike the Stilling test, the same designs appear in every edition of the test. Shortened versions of the Ishihara plates are available and a special test for children, or 'unlettered persons', has recently been introduced.

Munsell colours have superseded manufactured materials, such as wools and worsteds, for sorting tests. Munsell colours can be identified in the CIE system and test parameters can be accurately documented and reproduced. The use of a standard illuminant (source C) and a standardized test method ensures that the same procedures are adopted by examiners everywhere and that similar results can be obtained in different test locations.

Tests in current use

The Holmgren wool test (Sweden)

The Holmgren wool test originally consisted of a selection of 125 small skeins of wool but modern versions of the test have either 75 or 49 skeins.

There are three test skeins, yellow–green, rose (pink), and purple. The wools are placed in a heap on a neutral grey background and the person selects 'matching shades' for each test skein. There is no exact match for the test colour, the task is to select skeins which are the same hue but lighter or darker (i.e. different only in value and chroma). A numerical code is attached to each skein so that the examiner can identify correct and incorrect choices. A total viewing time of 1 min is recommended for the test and only pass or fail is recorded. The test has several disadvantages. Firstly, the three test colours are not the best ones to choose for accurate colour vision screening and, although colour confusions can be demonstrated, the pass/fail level is inconsistent. Secondly, people pick up the skeins in order to inspect them more carefully. This changes the angular subtends of the colours and they eventually get dirty. Finally, it is not possible to control the time taken to examine individual samples. In practice, a person takes longer to examine colours which are a close match and this helps some colour deficient people to obtain the correct result.

The unreliability of the Holmgren wool test is documented in the famous legal case involving the British merchant seaman John Trattles. Trattles went to sea in 1897 and passed examinations, including the Holmgren wool test, to obtain his second mate certificate in 1904. The following year he presented for his first mate certificate but failed the wool test and was required to surrender his second mate certificate. This he refused to do. A series of appeals and re-examinations followed. An examination with spectral colours confirmed that Trattles was colour deficient, but between 1904 and 1907 he had been given the Holmgren test on six separate occasions, passing on three and failing on three. A court of enquiry was set up, following a debate in the House of Lords, and Trattles was finally granted his first mate certificate in 1910. In spite of additional evidence from the Board of Trade that approximately one-third of people failing the Holmgren wool test passed on re-examination, the test is still in use and is recommended as a 'standard' test for some occupations. Furthermore, medical officers in industry sometimes copy the format of the Holmgren test and construct similar tests from manufactured items. These tests can only be effective if given by an experienced and knowledgeable examiner, and should not replace pseudoisochromatic plates for colour vision screening.

Pseudoisochromatic plates

Printed pseudoisochromatic plates are the most widely used screening tests for abnormal colour vision. Most tests are designed to detect congenital red–green colour deficiency and only a few tests include designs to detect tritan defects. Many different tests have been produced containing similar designs to the Ishihara plates. The colour quality varies and none are superior to Ishihara for colour vision screening. Several different tests have been produced in Japan and the US and others have been published

Table 7.1 Function of the principal pseudoisochromatic tests for screening, grading, and classifying defective colour vision

	Screening		Classifying		
	Red–Green	Tritan	Protan/ deutan	Tritan	Grading
Ishihara	Yes	No	Yes	No	No
American Optical Co. (HRR)	Yes	Yes	Yes	Tritan/ tetartan	Mild, moderate, and severe categories
Ohkuma	Yes	No	Yes	No	Mild, moderate, and severe categories
Tokyo Medical College	Yes	Yes	Yes	Yes	Slight and severe categories
SPP 1	Yes	No	Yes	No	No
SPP 2	Yes	Yes	No	Yes	No
Dvorine	Yes	No	Yes	No	No
F2 plate	Yes	Yes	No	Yes	No
Velhagen	Yes	Yes	Yes	Yes	No
Bostrom–Kugelberg	Yes	No	No	No	No
Lanthony	No	Yes	No	Yes	Mild, moderate, and severe
New City University tritan test	No	Yes	No	Yes	Slight and severe categories

in Germany, France, Sweden, Spain, the CIS, the Republic of Korea, and the Republic of China. Only some of these tests have been used outside their country of origin and only the Ishihara plates and the American Optical Co. (Hardy, Rand, and Rittler) plates have been used throughout the world. The principal pseudoisochromatic plate tests are listed in Table 7.1.

The Ishihara pseudoisochromatic plates (Japan)

The Ishihara pseudoisochromatic test is the most widely used screening test for red–green colour deficiency, and clinical trials have shown that it is the most efficient. The test is not designed to detect tritan defects. The Ishihara test has the most varied and complex design format of all pseudoisochromatic tests. The full version contains 38 plates. Twenty five plates contain numerals and thirteen plates contain pathways. An abridged version containing 24 plates is available in most editions and a concise version containing 14 plates was introduced in 1989 (Table 7.2). The 14 plate test differs from the other editions in that the designs are printed on a white background instead of on individual cards interleaved into black card. A new Ishihara test for 'unlettered persons' was introduced in Japan in 1970 but did not become available in the UK until 1990. The 'unlettered persons' test is intended for the examination of

Table 7.2 Selection of plates from the 38 plate version of the Ishihara plates included in the abbreviated and concise editions

	Numeral designs																								
	Intro.			Transformation									Vanishing						Hidden digit			Protan/deutan class.			
Standard 38 plates	1*	2	3	4	5	6†	7	8	9	10	11	12	13	14	15	16	17	18	19	20	21	22	23	24	25
Abbreviated 24 plates	1	2	–	3	–	4	5	6	7	–	8	–	9	10	11	12	13	14	–	15	–	16	17	–	–
Concise 14 plates	1	2	–	4	–	3	–	5	–	8	–	–	7	–	6	10	–	–	9	–	–	–	–	12	13

	Pathway designs												
	Protan/deutan class.		Hidden digit		Vanishing				Transformation				Intro.
Standard 38 plates	26‡	27	28	29	30	31	32	33	34	35	36	37	38
Abbreviated 24 plates	18	–	–	19	20	–	21	–	–	22	–	23	24
Concise 14 plates	14	–	–	–	–	11	–	–	–	–	–	–	–

* Most similar design to plates 1 and 2 of the Ishihara test for 'unlettered persons'.
† Most similar design to plates 3 and 4 of the Ishihara test for 'unlettered persons'.
‡ Plate 8 of the Ishihara 4 test for 'unlettered persons'.

NB: The plates included in the concise edition are in a different order and the numbers given refer to that order.

children between 4 and 6 years of age. There are eight plates, four plates containing geometric shapes and four plates containing simplified pathways.

The Ishihara test is intended to be viewed at two-thirds of a metre (at arm's length). This distance maintains the correct angular subtends of the dot matrix. The MacBeth easel lamp is an ideal illuminant. The book is placed in the tray underneath the light source so that illumination equivalent to source C is provided at an angle of 45° to the surface of the plate. A viewing time of about 4 s is allowed for each plate. Of the 25 plates containing numerals, 1 is for demonstration of the visual task, 20 are for red–green screening, and 4 are for protan/deutan classification. The plates are arranged as follows:

(a) Plate 1. An introductory plate which is seen correctly by all observers. This plate demonstrates the visual task and can be used to detect malingerers.

(b) Plates 2–9. Transformation designs. A number is seen by normal trichromats and a different number by colour deficient observers.

(c) Plates 10–17. Vanishing designs. A number can be seen by normal trichromats but cannot be seen by colour deficient people.

(d) Plates 18–21. Hidden digit designs. A number cannot be seen by normal trichromats but can be distinguished by most colour deficient observers. These designs are less efficient in demonstrating colour deficiency than the preceding ones.

(e) Plates 22–5. Classification designs to distinguish protan and deutan defects. These are vanishing designs consisting of neutral colours. Two numbers are contained in each plate. Only the right hand (blue–purple) number can be seen by protan observers and only the left hand (red–purple) number by deutan observers. A typical response is not always made. A response that neither number is visible suggests a severe red–green colour deficiency coupled with a high density of macular pigment. If both numbers are seen, and errors have been made on the screening plates, classification of the type of colour deficiency is made by ascertaining which of the two numbers looks clearer or brighter. The number seen less clearly indicates the type of colour deficiency.

In the 38 plate test, plates 26–38 contain pathways and are intended for use with nonverbal subjects. The examination begins at the back of the book. Plate 38 is the demonstration design, plates 34–7 are transformation designs, plates 30–3 are vanishing designs, plates 28–9 have hidden pathways, and plates 26–7 are for protan/deutan classification.

All the screening plates containing numerals except the hidden digit designs, are extremely effective. Dichromats and anomalous trichromats fail most of the plates and cannot be distinguished by the number of errors made. Only people with slight protan and deutan defects read some plates correctly and undue hesitation is often a sign of slight colour deficiency. If both protan/deutan classification numerals can be distinguished, on plates 22–5, colour deficiency is usually slight.

The serif designs of the numerals in the Ishihara plates leads to misreadings as some normal trichromats complete partial loops. For example, ‘5’ may be interpreted as ‘6’, or ‘3’ as ‘8’. Misreadings on vanishing designs should not be included as errors or as failure of the plate. However, completing partial loops on transformation designs gives an ambiguous result if the response is a composite of both the correct and ‘confusion’ numerals. This type of misinterpretation is sometimes referred to as a ‘partial error’ and can be made by both normal and colour deficient observers. Colour deficient people invariably record clear errors on other plates and so there is no uncertainty in the overall result. The large number of plates, in different design categories, assists the examiner to place ‘partial errors’ into the correct perspective. Abbreviating the test, to achieve more rapid colour vision screening, may therefore create difficulties unless plates which often produce ‘partial errors’ are avoided. Six plates are the minimum required for rapid screening. The ‘sensitivity’ of each plate of the Ishihara test has been determined from clinical trials and the six most efficient designs are plates 2, 3, 5, 9, 12, and 16. This selection includes four transformation designs and two vanishing designs. Unfortunately three of these plates do not appear in either the abbreviated or concise versions of the test and some designs, such as plates 4 and

13, which are particularly liable to misinterpretation, are included. The abbreviated and concise tests appear to have been compiled from the point of view of economy rather than to provide more accurate colour vision screening and if rapid screening is required it is better to select plates from the 38 plate edition.

The personal nature of the examination and slight differences in colour printing in different test editions makes it impossible to state categorically the maximum number of misreadings which can be allowed to normal trichromats and the minimum number of errors which constitute slight colour deficiency. Knowledge of the test method and the definition of 'misreading' and 'error' is needed. Results for the ninth and 1989 test editions are compared in Table 7.3 a,b. The colour printing in these two editions is noticeably different and they represent the possible limits of colour specification tolerance. The results show that normal trichromats often make one or two misreadings and occassionally as many as six misreadings. Colour deficient people rarely make as few as nine errors on the 1989 edition or seven errors on the ninth edition. The separation of the normal response and the responses of slightly colour deficient people, in terms of mistakes, is greater for the 1989 edition making it a better screening test. Most colour deficient people, dichromats, and anomalous trichromats, fail all the transformation and vanishing designs and it is not possible to estimate the severity of colour deficiency. Only some people with slight colour deficiency fail a fewer number of plates as shown in Table 7b. The same five colour deficients failed a small number of plates on both the ninth and 1989 editions of the test.

Explanatory notes on design and on test administration are supplied with the Ishihara test but these are printed in a separate paper booklet which frequently becomes separated from the test.

The American Optical (Hardy, Rand, and Rittler) plates (AO HRR plates) (US)

The HRR plates were first printed in 1954 and there have been two identical editions of the test. The plates are currently out of print but a new edition is being considered. The HRR test is designed to identify protan, deutan, tritan, and 'tetartan' defects and to grade their severity. The 'tetartan' designs are superfluous. The test consists of 24 plates containing symbols (circles, crosses, and triangles) and employs neutral colours which increase in saturation in successive plates of the test. All the designs are of the vanishing type. There are 4 introductory plates, 6 plates for colour vision screening, and 14 plates for grading the severity of protan, deutan, and tritan defects. These are arranged as follows.

(a) Plates 1–4 are introductory plates.

Screening plates

(b) Plates 5–8 are for detecting protan and deutan defects.

(c) Plates 9–10 are for detecting tritan and 'tetartan' defects.

If plates 5–10 are interpreted correctly, the person is considered to have normal colour vision and the test is concluded. If mistakes are made the remaining plates are given.

Table 7.3a Specificity and sensitivity of each plate of the 1989 and ninth editions of the Ishihara test

		Intro.			Transformation designs									Vanishing designs					Hidden digit designs			
		Plate 1	2	3	4	5	6	7	8	9	10	11	12	13	14	15	16	17	18	19	20	21
Ishihara	Specificity		99.8	99.8	93.4	97.2	98.7	97.2	99.8	92.6	100.0	100.0	89.4	89.4	99.8	100.0	98.7	80.0	100.0	100.0	100.0	100.0
1989	Sensitivity		93.0	93.0	93.0	100.0	93.0	93.0	89.7	100.0	89.7	89.7	96.6	89.7	93.1	93.1	96.6	100.0	17.0	31.0	59.0	31.0
	Efficiency		192.8	192.8	186.4	197.2	191.7	190.2	189.5	192.6	189.7	189.7	186.0	179.1	192.9	193.1	195.3	180.0	117.0	131.0	159.0	131.0
Ishihara	Specificity		100.0	100.0	97.4	99.8	99.5	98.4	99.8	94.0	99.6	99.6	84.1	77.5	99.5	100.0	97.9	91.5	100.0	100.0	100.0	100.0
ninth	Sensitivity		96.6	93.1	79.0	93.1	86.2	82.7	82.7	100.0	93.1	89.7	93.1	86.2	89.7	89.7	93.1	89.7	34.0	45.0	44.8	38.0
edition	Efficiency		196.6	193.1	176.4	192.9	185.7	181.1	182.5	194.0	192.7	189.3	177.2	163.7	189.2	189.7	191.0	181.2	134.0	145.0	144.8	138.0

Table 7.3b Number of misreadings by 471 normal trichromats (N) and number of errors made by 29 red–green colour deficient observers (CD) on the 16 screening plates of the Ishihara test (Transformation and vanishing numeral designs only)

No. of misreadings	Ishihara 1989		Ishihara ninth edition	
or errors	N	CD	N	CD
0	259	–	280	–
1	141	–	127	–
2	41	–	45	–
3	21	–	14	–
4	6	–	2	–
5	2	–	1	–
6	1	–	2	–
7	–	–	–	1
8	–	–	–	1
9	–	2	–	1
10	–	–	–	–
11	–	1	–	–
12	–	1	–	1
13	–	1	–	–
14	–	–	–	1
15	–	–	–	–
16	–	24	–	24

Short rules denote demarcation between the number of misreadings made by normal trichromats and the number of errors made by colour deficient observers.

Grading plates

(d) Plates 11–20 are for grading the severity of protan and deutan defects. Three categories are distinguished: slight (plates 11–14), moderate (plates 15–18), and severe (plates 19–20).

(e) Plates 21–4 are for grading the severity of tritan and 'tetartan' defects. Two categories are distinguished: moderate (plates 21–2) and severe (plates 23–4).

The four red–green screening plates are low threshold designs and many people with normal colour vision fail to interpret the first two designs correctly (Belcher *et al.* 1958). Some colour deficient people obtain the same result and the test is therefore not suitable for accurate colour vision screening. The HRR test is valuable for its grading function and for the detection of tritan defects. Only moderate or severe tritan defects are identified. Although the grading plates are effective, it is not possible to distinguish dichromats and severe anomalous trichromats from the test results. Some dichromats are classified in the moderate category and some anomalous trichromats in the severe category. Many examiners change the order of the plates and present the tritan screening plates before the low threshold red–green screening plates; this provides a better introduction to the test. The plates can be presented in reverse order if severe colour deficiency is suspected. This has the advantage of working from plates which are easy to interpret to those which are more difficult instead of vice versa.

A score sheet and test instructions are provided.

The Ishihara and HRR plates are often used together in a clinical examination. The Ishihara plates are used for red–green screening and the HRR plates for estimating the severity of colour deficiency and for tritan screening.

Four other pseudoisochromatic plates have been published in Japan and three other tests in the US. None of these tests have been used as widely as the Ishihara plates or the HRR plates and some tests are much less effective.

The Ohkuma plates (Japan)

The format of the Ohkuma test is similar to the Ishihara test but the contained figure is a Landolt C. The test was first introduced in 1973 and is for red–green colour deficiency only. There are 14 plates, 1 plate is for demonstration, 7 are for screening, and 6 for grading and classifying protan and deutan defects. Four of the screening plates are of the transformation type, two are vanishing designs and one has a hidden digit format. In the transformation designs the gap in the 'C' should be seen in different positions by colour normal and colour deficient observers. The six protan/deutan classification and grading plates are in two series of three plates for each type of defect. The size of the 'C' and the colour saturation increases in each series. The Ohkuma test introduced several new design concepts. However, the transformation designs are ambiguous and normal trichromats, and some colour deficient people, can see both breaks in the 'C'. This results in poor screening efficiency. The most reliable screening plates are plates 6 and 7 (vanishing designs) and plate 8 (hidden digit design).

The Tokyo Medical College test (TMC test) (Japan)

The TMC test is constructed by placing a white grid on top of painted colours. There are thirteen plates. The size of the dot matrix is the smallest used in pseudoisochromatic plates and the gaps between the dots is relatively large. Visual acuity better than 6/12 is needed to distinguish these designs and this limits application of the test for acquired colour deficiency. No introductory plate is provided. There are five red–green screening plates and two tritan screening plates. Six plates are for protan/deutan classification and are also intended to estimate the severity of colour deficiency. All the plates contain numerals and are of the vanishing type. The red–green screening plates are reliable for detecting congenital colour deficiency but the tritan screening designs are less effective. Classification and grading protan and deutan defects is not correctly realized in the TMC test. Many colour deficient people cannot see either of the numerals on the classification plates.

A score sheet and instructions on administration are provided with the test.

The standard pseudoisochromatic plates first edition (SPP 1 test) (Japan)

The SPP 1 test is intended to screen and classify congenital red–green colour deficiency. Use of a regular dot matrix restricts the design of the contained digits to the format used in computers and visual display units. Some people find it difficult to interpret this format and examples, which are provided, have to be shown before the test begins. There are 19 plates. There are 4 demonstration plates, 10 screening plates consisting of vanishing designs, and 5 protan/deutan classification plates. The screening plates effectively distinguish between normal trichromats and people with congenital red–green colour deficiency but the protan/deutan classification is not correctly realized. In addition, tritanopes cannot distinguish some of the designs intended to screen for red–green colour deficiency and cannot interpret plates 2 and 3 of the demonstration plates. These results appear to be unintentional. The test instructions and score sheet are confusing. It is unnecessary for the observer to comment on the relative clarity of the paired digits contained in the screening plates as recommended.

The standard pseudoisochromatic plates second edition (SPP 2 test) (Japan)

The SPP 2 test is intended for the examination of acquired colour vision defects. The format is similar to that of the SPP 1 test. There are 12 plates in all. Two plates are for demonstration. Each of the remaining 10 screening plates contain pairs of digits. These include two more digits for demonstration. There are 4 digits for detecting red–green defects (type 1 and type 2 acquired defects), 12 digits for detecting acquired type 3 (tritan) defects, and 2 digits for detecting scotopic vision (rod monochromats). The colour differences used on plate 3 are very small and the digit to detect tritan defects cannot be seen by most normal trichromats. Failure to see this digit should not be recorded as an error indicating a tritan type of deficiency. The SPP 2 test is a good screening test for congenital red–green colour deficiency and only slight deuteranomalous trichromats make no errors and pass the test. No classification of protan and deutan, or of type 1 and type 2 acquired defects, is attempted. The designs intended to distinguish scotopic vision are failed by protans. Patients with acquired type 3 defects often give a mixed red–green and tritan diagnosis even when other tests show a clear type 3 deficiency. This result is due to the small colour differences used for the red–green designs. Although, some of the intended features of the test are not realized, the SPP 2 test has considerable potential for detecting both congenital and acquired colour deficiency.

A score sheet is provided but the instructions for administering the test are confusing. It is not necessary to compare the clarity of the paired digits on every plate as suggested in the test manual and the test should be given in the normal way.

The Dvorine pseudoisochromatic plates (US)

The Dvorine test is a standard screening test for congenital red–green colour deficiency and is widely used in the US. Screening accuracy is similar to that of the Ishihara test. The Dvorine test originally consisted of two volumes containing 120 plates and the designs have been revised and reprinted many times. The fourth edition of the test was printed in 1953 and consists of 23 plates. Fifteen plates contain numerals and eight plates have pathways for the examination of nonverbal subjects. The printed colours are similar to those of the Ishihara test. Of the numeral plates, plate 1 is for demonstration and 12 plates are for red–green screening, 2 plates (plates 5 and 6) are for protan/deutan classification. The efficiency of the classification plates is low. All the plates consist of vanishing designs with the consequence that many colour deficient observers cannot interpret any of the plates except the introductory design. The test manual states the minimum number of 'errors' which result in failure of the test but does not address the problem of how to interpret misreadings. Misreadings on vanishing designs are not 'errors', whereas undue hesitation and failure to see, some, or all, of the designs identifies colour deficiency.

The Dvorine test contains a colour naming test in addition to pseudoisochromatic plates. This consists of eight saturated and eight desaturated colours mounted on a cardboard wheel. Each colour is shown singly and has to be named by the observer. This examination is of doubtful value and no instructions are given on how to administer the test or interpret the results.

The American optical colour vision test (US)

The American optical test was originally printed in 1940 and contains 15 red–green screening plates copied from European and Japanese tests. The colour quality is only moderate and the plates are not as effective as the originals.

The Farnsworth F2 plate (US)

Dean Farnsworth designed the tritan screening plate (the F1 plate) used in the *Picture Post* survey. The F1 plate was a transformation design containing two numerals based on the design of plate 6 of the Ishihara test. Farnsworth's second design, the F2 plate, consists of two overlapping squares, one yellow–green and the other blue, on a background matrix of purple dots. Normal trichromats and tritanomalous trichromats see both squares. Normal trichromats describe the yellow–green square as clearer, or more prominent, and tritanomalous trichromats see the blue square more clearly. Tritanopes are intended to see the blue square only. An apparently unintentional bonus is that protans and deutans cannot see the yellow–green square. The F2 plate is therefore a possible all-purpose colour vision screening test and is actually more efficient for detecting red–green defects than it is for detecting tritan defects. The plate is only partially successful in identifying congenital tritan defects and is not reliable in detecting acquired type 3 (tritan) colour deficiency. Individual responses to the design have to be carefully interpreted. The observer has to say how many squares there are and point to the clearer square without touching the design.

The F2 plate is not available commercially but can be constructed from Munsell samples (Taylor 1975). The background dots are Munsell 5P 6/4 and 7/4, the green square is mainly 2.5GY 6/4 and 7/4, and the blue square is principally 7.5BP 6/4 and 7/4.

The Velhagen pseudoisochromatic plates (Germany)

The Velhagen test is derived from the original Stilling plates. The twenty-sixth edition contains a combination of numerals and upper-case letters. There are 30 designs but the test can be abbreviated to about 17 plates without losing screening efficiency (Frey 1962). Earlier editions of the test are still in use and the contained designs are not quite the same in each edition. Either one or two introductory plates are provided, and there are a large number of red–green screening plates of the vanishing type and two or more screening plates intended for tritan defects. Classification plates for protan, deutan, and tritan defects are included. Some editions in this series contain designs for the examination of nonverbal subjects and some do not. The test is haphazardly arranged and there is no uniform style for either the dot sizes or the figures. Two designs to test simultaneous colour contrast and one to detect 'shortening of the red end of the spectrum' are included.

The latest edition of the test has been revised by Broschmann. The test instructions are in German but no score sheet is provided. This means that an examiner has to carefully familiarize himself with the test before using it and knowledge of isochromatic confusion colours is needed in order to understand the intended function of each design.

The Bostrom–Kugelberg pseudoisochromatic plates (Sweden)

The Bostrom and later the Bostrom–Kugelberg pseudoisochromatic plates have appeared in several different editions. The current Bostrom–Kugelberg test was introduced in 1974 and consists of 20 plates. There are 3 demonstration plates, 15 red–green screening plates containing vanishing designs, and 2 plates containing pathways for the examination of nonverbal subjects. Desaturated colours having small colour differences are used for all the plates and the figures have a very narrow stroke width. An illuminant giving more than 500 lx is recommended and visual acuity of 6/9 or better is needed to complete the test.

The Lanthony tritan album (France)

The Lanthony tritan album contains a demonstration plate and 5 plates for detecting and grading congenital tritan defects and acquired type 3 defects. The plate design consists of a square background of grey dots. One corner of the square contains blue–purple dots. The blue dots decrease in saturation from plates 1–5 to give a series of vanishing designs with different colour difference steps. The observer has to identify the corner of the square which appears coloured. The efficiency of the test for detecting congenital tritan defects has not been evaluated but only severe acquired type 3 defects can be detected by the least saturated design (plate 5). Patients with slight or moderate type 3 defects pass. These results show that neither the screening nor the intended grading function of the test is correctly realized.

The City University tritan test (UK)

The City University tritan test consists of 5 plates. Three plates are for tritan screening and two plates are to distinguish severe tritan defects. The test is intended to supplement the Ishihara test to provide comprehensive colour vision screening. There is no introductory design. All 5 plates are of the vanishing type and contain circle, cross, and triangle symbols. The colour difference steps are deliberately chosen to be fairly large in order to reduce the number of false positive results. The plates successfully identify congenital tritan defects and have been used extensively in the study of acquired colour deficiency in diabetic patients.

Miscellaneous plate tests

The FVS plates (Japan)

The FVS (Francois, Verriest, Seki) plates are designed to identify typical rod monochromatism. There are 3 plates, 1 plate is to demonstrate the visual task and 2 plates are for diagnosis. Each design contains a coloured cross in a matrix of coloured dots. The luminance contrast of the colours is selected so that the designs cannot be seen by people with scotopic relative luminous efficiency. All other observers are able to distinguish the design.

The Berson plates (US)

The Berson pseudoisochromatic plates are designed to distinguish rare cases of typical and atypical ('blue cone') monochromats so that appropriate genetic counselling can be given without the need for electrodiagnostic tests. The plates are not available commercially but can be constructed from Munsell colours (Berson *et al.* 1983). There are 6 pseudoisochromatic designs, 2 plates are to demonstrate the visual task and 4 plates are for diagnosis. All the plates have vanishing designs containing arrows. The test is designed to identify different luminance contrast relationships in typical and atypical achromatopsia. However this can be examined in more detail with the Sloan achromatopsia test.

Computer controlled displays

A format similar to the HRR plates has been adopted in a prototype computer display test. The background matrix is composed of grey spots and the contained figure is a Landolt C. The observer presses one of four buttons corresponding with the position of the gap in the 'C'. The figure is presented in a series of saturation steps for each of the protan, deutan, and tritan neutral colours. The presentation time is limited to a few seconds. The test successfully classifies and grades different types of colour deficiency but, like the HRR plates, is not completely reliable as a screening test and cannot distinguish dichromats and anomalous trichromats.

Other pseudoisochromatic tests

The Polack pseudoisochromatic test (France) and the Rabkin pseudoisochromatic plates (CIS) have both appeared in a number of editions but are no longer available. The Polack test consisted of approximately 50 plates for detecting red–green deficiency but could be abbreviated to the best 14 plates if desired. Some editions of the Rabkin plates contain pictures and printed designs representing possible colour matches which might be achieved in an anomaloscope examination, as well as pseudoisochromatic designs.

Pseudoisochromatic tests containing numerals are currently published in Korea (the Hahn pseudoisochromatic plates) and tests containing Chinese characters are published in China.

Colour vision tests in multiphasic visual screening instruments

A number of visual screening instruments, such as the Titmus vision tester, the Keystone ophthalmic telebinocular, and the Bausch and Lomb orthorator, contain colour slides or photographic reproductions of pseudoisochromatic designs. Visual screening instruments are not recommended for colour vision screening. The colour quality of these reproductions differs from the originals and screening accuracy is reduced, sometimes drastically. The Rodenstock vision screener contains a 'colour wheel test' which mimics an anomaloscope examination. The accuracy of this test has not been determined.

Hue discrimination or arrangement tests

Farnsworth–Munsell tests were developed from arrangement tests, containing manufactured colour samples used to select industrial workers in the 1920s.

The Farnsworth D15 test

The prototype D15 test, the dichotomous test, was described by Dean Farnsworth in 1943. The standard D15 test divides people into two groups. The first group consists of people with normal colour vision and slight colour deficiency; the second group consists of people with moderate and severe colour deficiency. Farnsworth considered that the former would be able to use industrial colour codes safely and that the latter would be unsafe. Clinical trials show that about 5 per cent of men fail the D15 test compared to the known incidence of 8 per cent red–green defects. A single error of two steps or more results in failure of the test. The colour arrangement made by the observer is drawn on a circular diagram representing the hues. Isochromatic errors give rise to lines which cross the diagram and show that colours from opposite sides of the hue circle have been placed next to each other in the arrangement. Typical results are obtained in congenital protan, deutan, and tritan colour deficiency. The colour difference steps are not uniform across the hue circle and isochromatic confusions equivalent to smaller difference steps are made more frequently. Two grades of deficiency (moderate or severe) can be identified from the number of isochromatic confusions made (Fig. 7.1a).

Some examiners include an extra grey cap (Munsell value 5) to demonstrate neutral colours. This is kept in reserve until the arrangement has been completed and isochromatic confusions have been demonstrated. The observer is then asked to include the grey cap in the sequence or to select matching samples.

Monochromats may be able to place the colours in an achromatic lightness scale which corresponds with their relative luminous efficiency. Typical results have been reported for rod monochromats (Fig. 7.1b).

The standard D15 test has Munsell colours with value 5 and chroma 4 and is very efficient for grading the severity of colour deficiency. Two other D15 tests containing the same Munsell hues but different chroma and values have been used clinically. These tests have different pass/fail levels and are used to grade the severity of colour deficiency in more detail. A D15 panel having Munsell value 5 and chroma 2 (sometimes called the Adam's test) is a useful second test if the level of illumination is 200–650 lx. The desaturated D15 test introduced by Lanthony has Munsell value 8 and chroma 2. This test is often referred to as the desaturated D15 test or the Lanthony D15 test. The Lanthony test is intended for colour vision screening but a large number of false positive results are obtained unless very high illumination levels, over 1000 lx, are used. The test cannot be used if the illumination is less than 500 lx.

D15 results are evaluated by visual inspection of the score sheet. However, several methods for calculating a numerical score have been proposed. The simplest method is to count the number of isochromatic 'crossovers' and their size in terms of cap displacement. A scoring method derived from colour difference steps has been developed by Bowman (Bowman 1982). The colour difference step between adjacent hues in the arrangement is calculated, using either the CIELUV or CIELAB formula, and a total error score obtained from the sum of the individual differences. However the derived total error scores have rather large numerical values and are difficult to relate to actual test performance. Tritanopes obtain much lower error scores than red–green dichromats because fewer isochromatic confusions can be made. Vector analysis can also be used to express isochromatic confusions numerically and to compute error scores from colour difference steps.

A test manual is provided which gives details of the test design but the clinical use of the D15 test is not fully explained.

Derivatives of the D15 test

The H16 test

The H16 test is similar to the D15 panel but contains different Munsell colours at a higher level of saturation (Table 7.4). The colours are derived from an incomplete hue circle and have Munsell value 5 and chroma of either 6 or 8. People who fail the H16 test are almost invariably red–green dichromats but some dichromats can pass the test and some severe anomalous trichromats fail. The H16 test is especially useful if a spectral

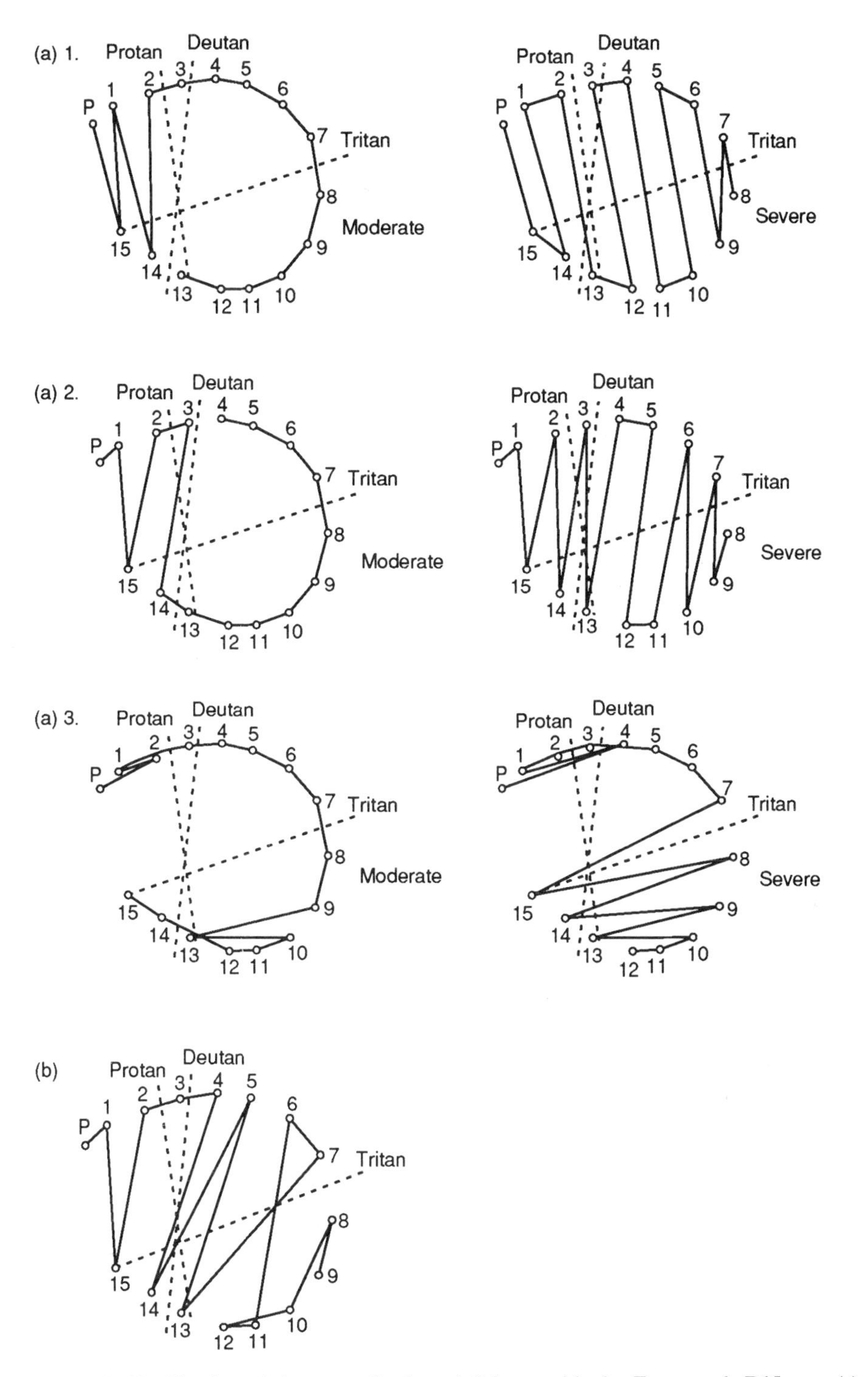

Fig. 7.1 Classification of the type of colour deficiency with the Farnsworth D15 test. (a) Protan, deutan, and tritan defects. 1. Moderate and severe protan defects. 2. Moderate and severe deutan defects. 3. Moderate and severe tritan defects. (b) Typical 'rod' monochromatism. The arrangement represents a lightness scale not isochromatic colour confusions.

Table 7.4 Munsell notation and CIE chromaticity coordinates for the Farnsworth D15 test and the Paulson H16 test

	D15 test Munsell notation		CIE x	CIE y	H16 test Munsell notation		CIE x	CIE y
Pilot	10.0B	5/6	0.228	0.254	2.5YR	5/10	0.504	0.383
1	5.0B	5/4	0.235	0.277	7.5R	5/10	0.496	0.341
2	10.0BG	5/4	0.247	0.301	5.0R	5/8	0.469	0.318
3	5.0BG	5/4	0.254	0.322	2.5R	5/8	0.431	0.304
4	10.0G	5/4	0.264	0.346	7.5RP	5/8	0.398	0.288
5	5.0G	5/4	0.278	0.375	5.0RP	5/8	0.365	0.268
6	10.0GY	5/4	0.312	0.397	10.0P	5/8	0.334	0.246
7	5.0GY	5/4	0.350	0.412	7.5P	5/8	0.313	0.225
8	5.0Y	5/4	0.390	0.406	2.5P	5/6	0.275	0.228
9	10.0YR	5/4	0.407	0.388	5.0B	5/8	0.230	0.236
10	2.5YR	5/4	0.412	0.351	5.0B	5/6	0.211	0.263
11	7.5R	5/4	0.397	0.330	10.0BG	5/6	0.217	0.289
12	2.5R	5/4	0.376	0.312	5.0BG	5/6	0.230	0.327
13	5.0RP	5/4	0.343	0.293	10.0G	5/6	0.238	0.364
14	10.0P	5/4	0.326	0.276	5.0G	5/6	0.269	0.391
15	5.0P	5/4	0.295	0.261	2.5G	5/6	0.289	0.421
16	–		–	–	10.0GY	5/6	0.316	0.454

From Paulson 1973.

anomaloscope is not available and dichromats or others with severe colour deficiency must be identified. The test lacks the necessary colours to evaluate tritan defects.

Paired D15 tests

The use of two or more D15 tests, with different saturation levels, can yield additional information about the severity of colour deficiency and can confirm the results obtained with pseudoisochromatic grading tests. If the standard D15 test is passed—a desaturated test is given. If the standard test is failed—The H16 test is given. Colour deficient people who complete the standard D15 and the desaturated Munsell 5/2 panel without error have very slight colour deficiency. People who fail the standard D15 and the H16 test have severe colour deficiency and are usually dichromats. The Lanthony D15 test does not provide more information than the 5/2 panel and classification of the type of deficiency may be less clear (Fig. 7.2).

The new color test

The new colour test was designed by Lanthony in 1975. The test contains 70 Munsell samples. Considerable expertise is needed to administer the

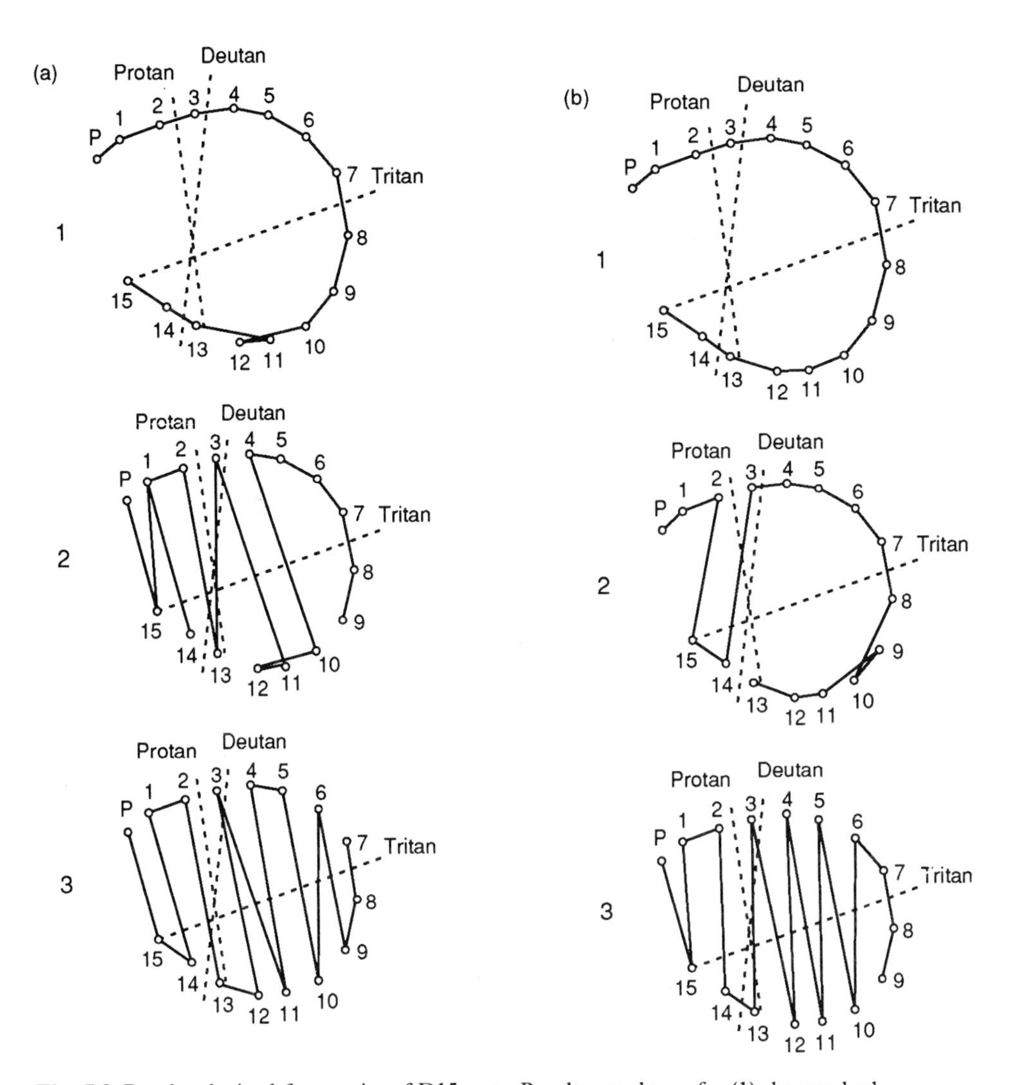

Fig. 7.2 Results obtained for a series of D15 tests. Results are shown for (1) the standard D15 test (Munsell 5/4), (2) a test with Munsell 5/2, and (3) the Lanthony desaturated test (Munsell 8/2). Both observers pass the standard D15, but the type of colour deficiency is correctly classified with the second test. More errors are made with the desaturated test (Munsell 8/2) and the type of colour deficiency is not easy to determine. (a) Protanomalous trichromatism. The observer obtained a protan axis of confusion with the F–M 100 hue test and an error score of 148. The range of match on the Nagel anomaloscope was 40–60 on the red/green mixture scale. (b) Deuteranomalous trichromatism. The observer obtained a deutan axis of confusion with the F–M 100 hue test and an error score of 96. The range of match on the Nagel anomaloscope was from 0 to 45 on the red/green mixture scale.

test effectively and no guidance is given in the test manual. The colours are divided into four series of 15 colours having Munsell value 6 and chroma of 2, 4, 6, and 8 respectively. This is equivalent to four D15 tests. In addition there are 10 grey caps, representing a lightness scale, which are used to determine neutral colours. The test is performed in two steps for each chroma series. The first step is a separation, or screening, phase in which the the colour caps are separated from the grey caps. The second step is a classification, or diagnostic, phase in which the colour caps are arranged in a colour order and the grey caps placed in lightness scale. This procedure is long and complicated. The test procedure can be simplified by giving the 6/4 series first and then proceeding in a logical manner. If the 6/4 series is failed the 6/8 series is given next and if this is passed the 6/6 series is given. If the 6/4 series is arranged correctly the 6/2 series is given. No reference colour is provided for the colour arrangement phase and it is easier for the observer if the examiner selects one. The examiner should also explain that it is not necessary to complete the hue circle by joining the first and last caps of the order. The results are plotted graphically and an error score calculated. Scoring is made easier if the examiner adds numbers to the colour notations designating the caps.

The new color test is not effective for colour vision screening. The majority of colour deficient observers complete the grey separation part of the test successfully and no neutral colours are identified. When the grey separation is incorrect it is assumed that grey caps will be included with the colours but the reverse can occur and the observer is left with fewer than 15 coloured caps to arrange in sequence. All 15 caps must be used in the classification part of the test so the examiner must correct the separation before proceeding. The following grades of severity are distinguished. Colour deficient observers either pass the test (slight colour deficiency), fail the chroma 2 series (moderate colour deficiency), of fail the chroma 8 series (severe colour deficiency).

The new color test is primarily intended for examining neutral zones in acquired colour deficiency. However the test has not been used extensively because of its complexity and because of inadequacies in the instruction manual.

The City University test

The City University test (TCU test) is derived from the D15 panel and contains the same Munsell colours together with grey. There have been two different editions of the test and only the second edition is currently available. Although the second edition contains designs with desaturated colours these are not effective in identifying a separate category of colour deficiency and both editions of the test have the same function. The TCU test is in book form and contains 10 plates. Each plate displays a central colour and four peripheral colours. The observer must select the peripheral colour which looks most similar to the central colour. Three colours are typical isochromatic confusions for protan, deutan, and tritan deficiency. The fourth colour is an adjacent colour in the D15 sequence and is the normal preference. The City University test is similar to the D15 test in that it identifies about 5 per cent of men who have red–green colour deficiency, but the classification of congenital protan and deutan defects is imprecise. Poor classification results from the limited choice of confusion colours. Several plates frequently give the incorrect classification compared to that shown on the score sheet (Fig. 7.3). Although the correct classification can be obtained from the most efficient designs, or

	Diagnosis		
Plate	Protan	Deutan	Tritan
1	••• ←		•
2	••	•••	
3	•	••	
4	••	•••	•
5	• ←		•
6		•	•
7	→	••	•
8	••	•••	•
9		•	•
10	→	••	•

Fig. 7.3 Classification of colour deficiency with the City University test (second edition). The number of dots indicates the efficiency of each plate in each classification category. ••• = Good classification efficiency; •• = moderate classification efficiency; • = slight classification efficiency. No dot indicates poor classification in this category. Arrows indicate plates which give incorrect classifications, for example, when protans respond as deutans.

from the majority of the results, it is best to consider that the test identifies people with moderate or severe red–green colour deficiency and to use other tests for protan/deutan classification. The test manual does not give clear instructions on the appropriate use of the test or on the interpretion of the results. The TCU test is a grading test not a screening test and 'mixed' defects do not occur as suggested in the test manual. In spite of these limitations, the TCU test is useful for detecting moderate or severe colour deficiency when a format other than the D15 is required.

The Farnsworth–Munsell 100 hue test (F–M 100 hue)

The prototype F–M 100 hue test was described by Farnsworth in 1943. The test examines hue discrimination ability and was originally intended for use in vocational guidance. The F–M 100 hue test is not a reliable screening test and only people with moderate or severe colour deficiency are identified. Hue discrimination ability is estimated from the total error score and the type of colour deficiency is determined from a graphical representation of the results. Characteristic F–M 100 hue plots for people with congenital protan, deutan, and tritan defects show concentrations of errors in two well-defined positions which are nearly opposite in the polar diagram representing the circle of hues (Fig. 7.4). These positions occur where isochromatic zones are tangential to the hue circle. The combined effect is of an axis of confusion centred around particular caps (Table 7.5). The axis of confusion is more prominent in severe colour deficiency leading to a higher error score. Protan and deutan error zones overlap slightly and classification of the type of colour deficiency is sometimes difficult to determine from the cluster of errors. Monopolar distributions of errors are sometimes found. Poor general hue discrimination is shown by random errors without an axis of confusion (Fig. 7.5). Due

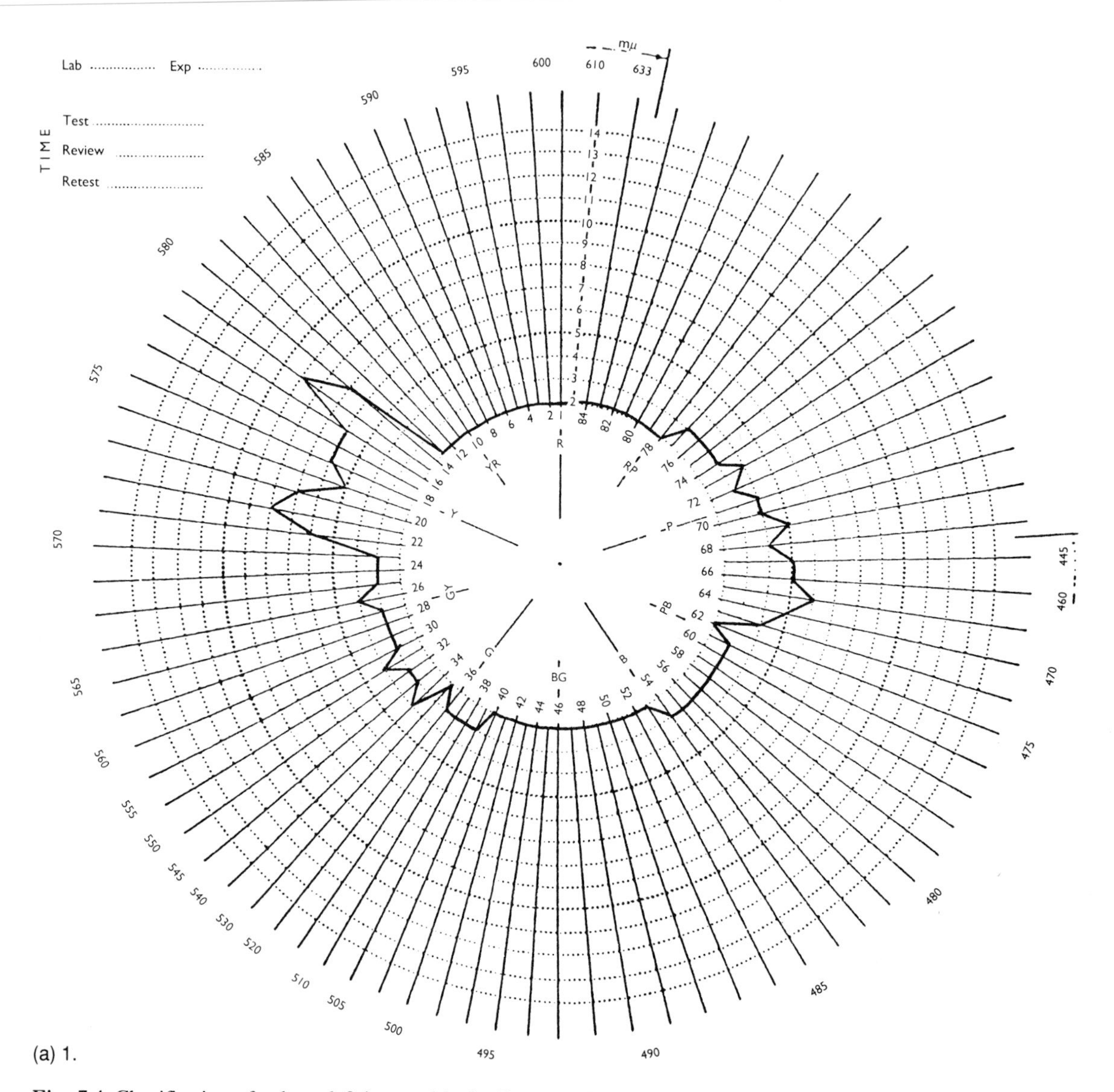

(a) 1.

Fig. 7.4 Classification of colour deficiency with the Farnsworth–Munsell 100 hue test. Typical axes of confusion are shown in protan, deutan, and tritan defects. (a) Protan defects. 1. Moderate protan axis (error score 100). 2. Severe protan axis (error score 192). (b) Deutan defects. 1. Monopolar deutan axis (error score 36). 2. Moderate deutan axis (error score 120). 3. Severe deutan axis (error score 304). (c) Tritan defects. 1. Moderate tritan axis (error score 140). 2. Severe tritan axis (error score 244).

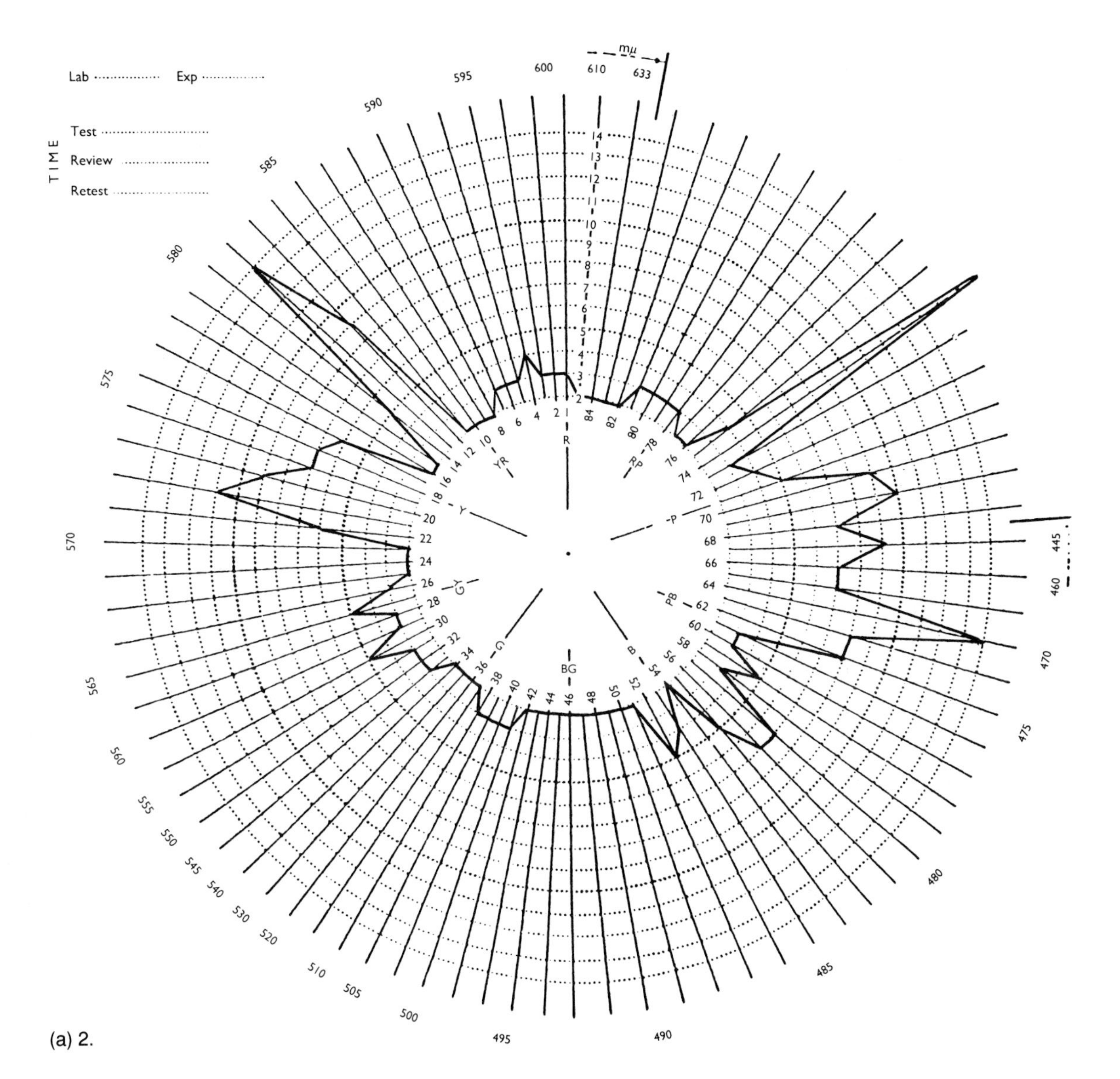

(a) 2.

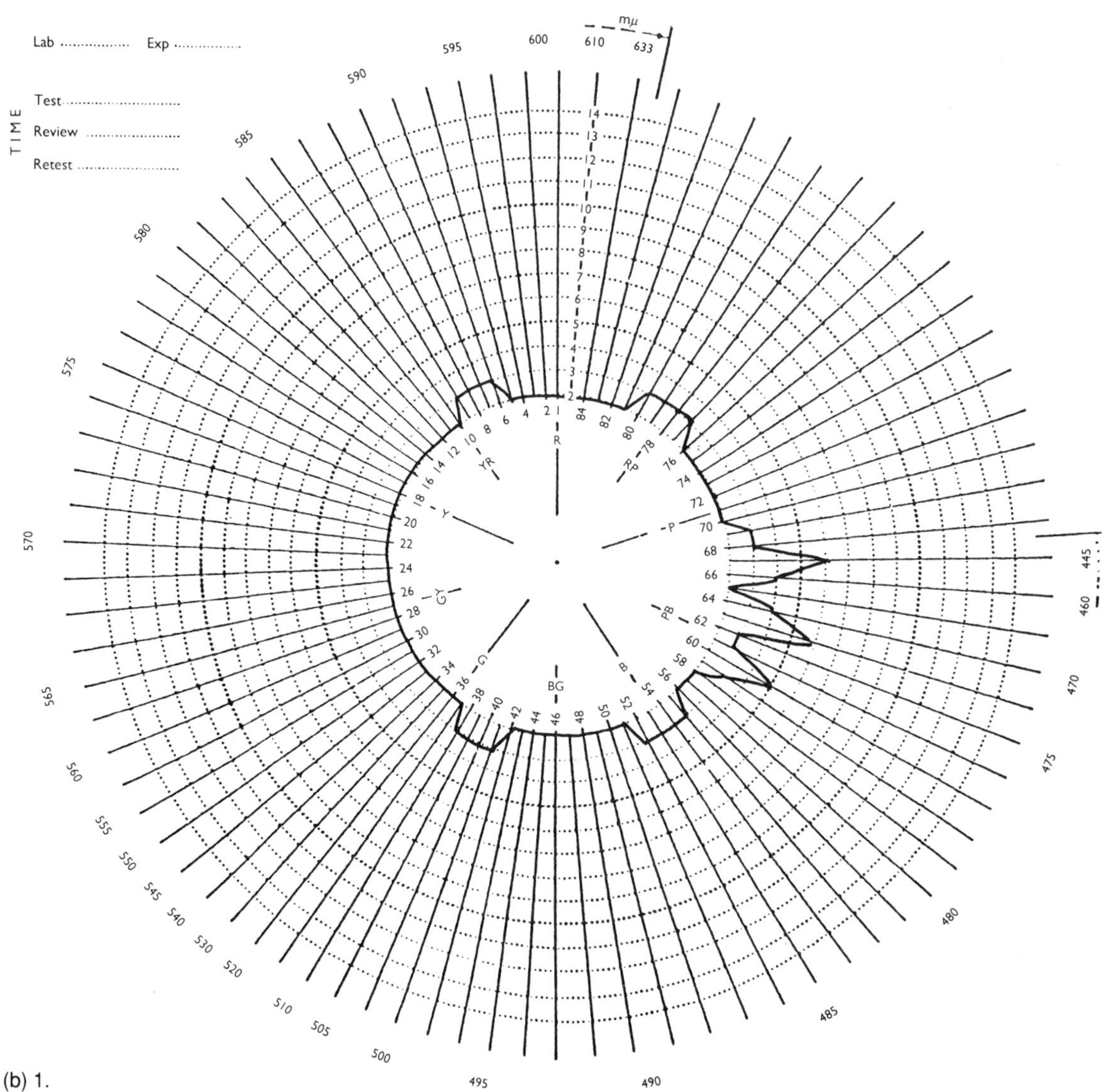
Lab .... Exp ....
TIME
Test ....
Review ....
Retest ....
mμ
633
610
600
595
590
585
580
575
570
565
560
555
550
545
540
530
520
510
505
500
495
490
485
480
475
470
460
445
R
YR
Y
GY
G
BG
B
PB
P
RP
1 2 3 4 5 6 7 8 9 10 11 12 13 14
2 4 6 8 10 12 14 16 18 20 22 24 26 28 30 32 34 36 38 40 42 44 46 48 50 52 54 56 58 60 62 64 66 68 70 72 74 76 78 80 82 84

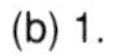
(b) 1.

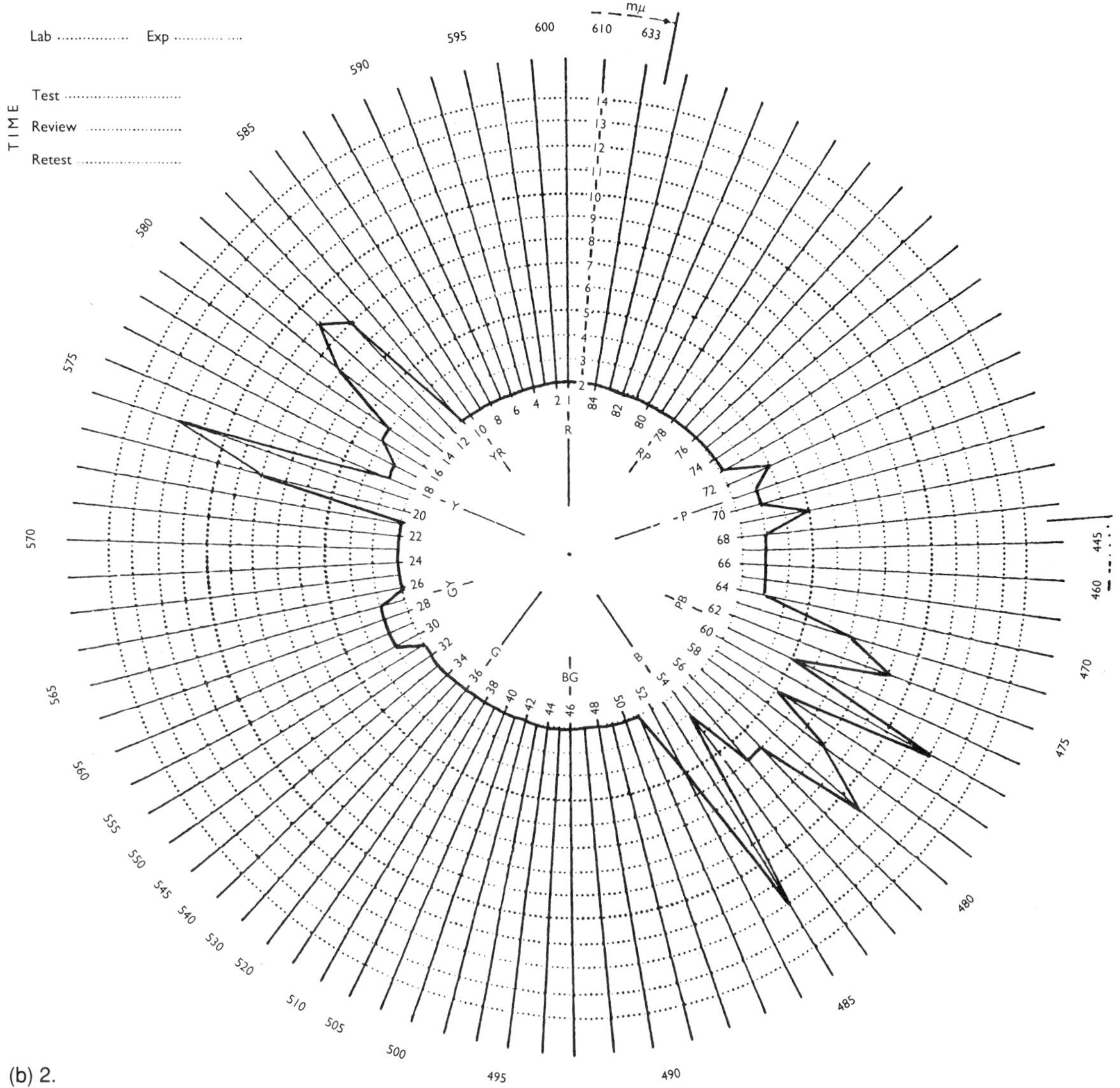
Lab .... Exp ....
TIME
Test ....
Review ....
Retest ....
mμ
600
610
633
595
590
585
580
575
570
565
560
555
550
545
540
530
520
510
505
500
495
490
485
480
475
470
460
445
R
YR
Y
GY
G
BG
B
PB
P
RP

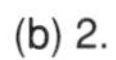
(b) 2.

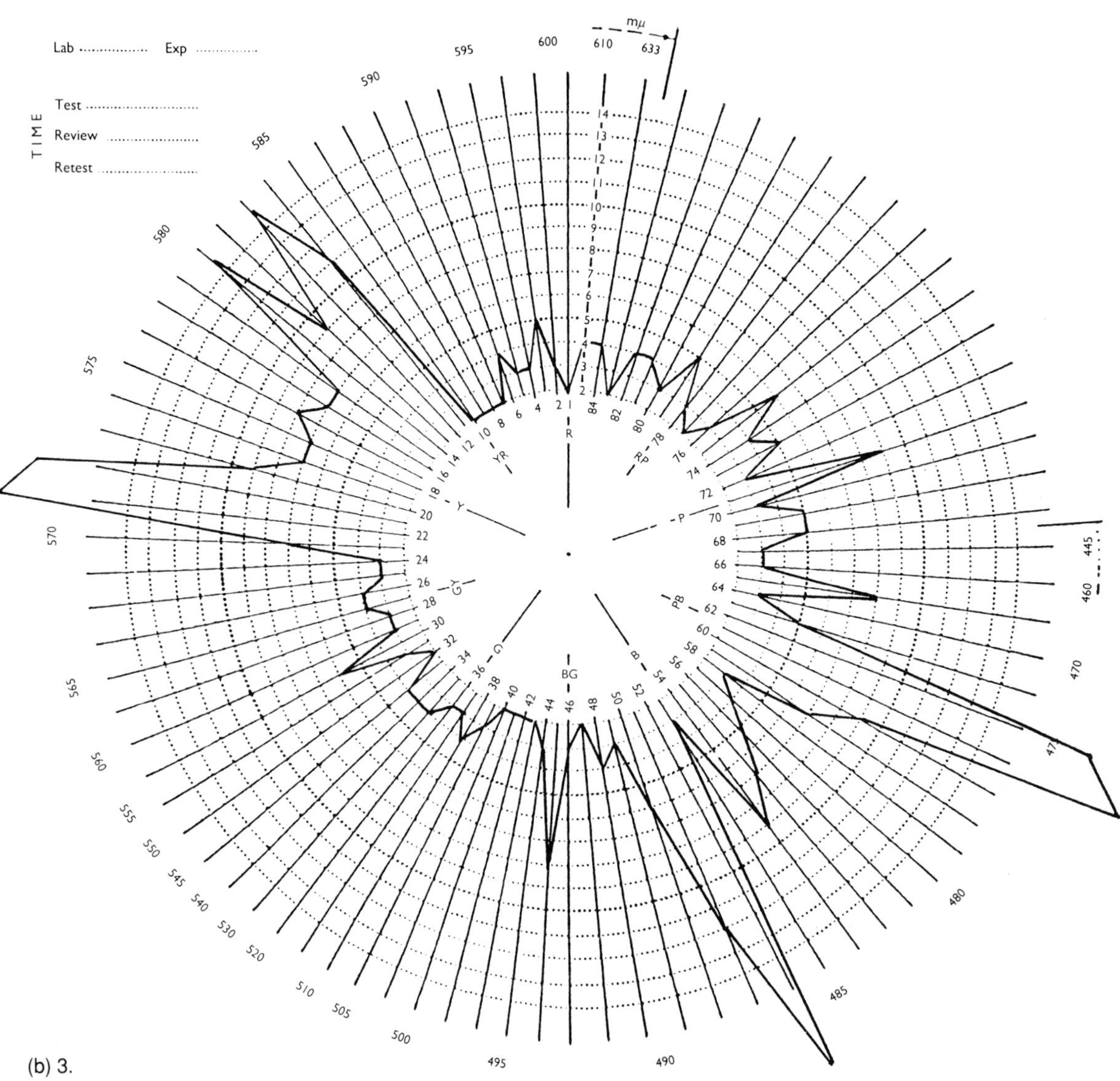

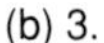
(b) 3.

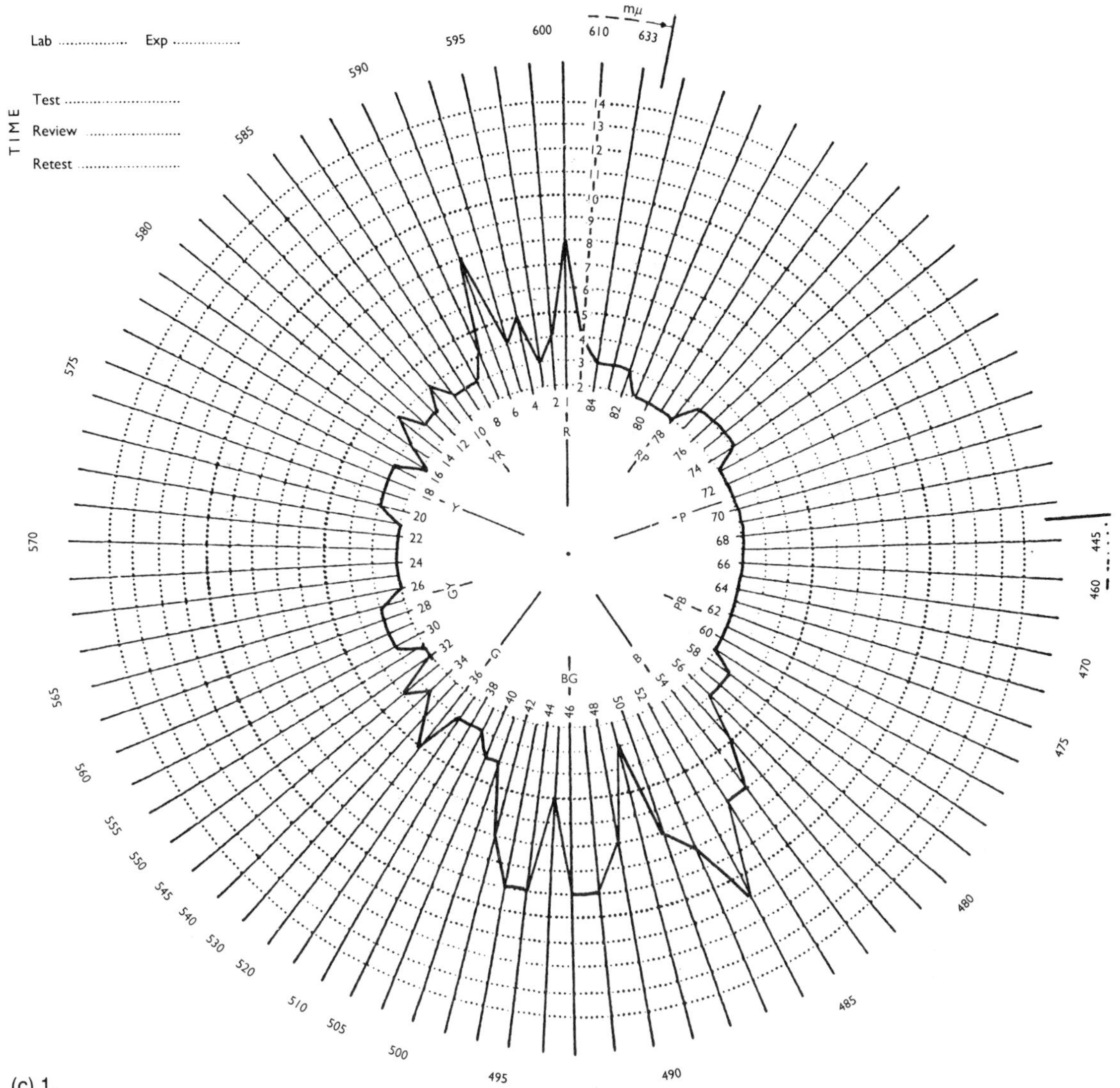

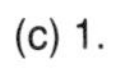
(c) 1.

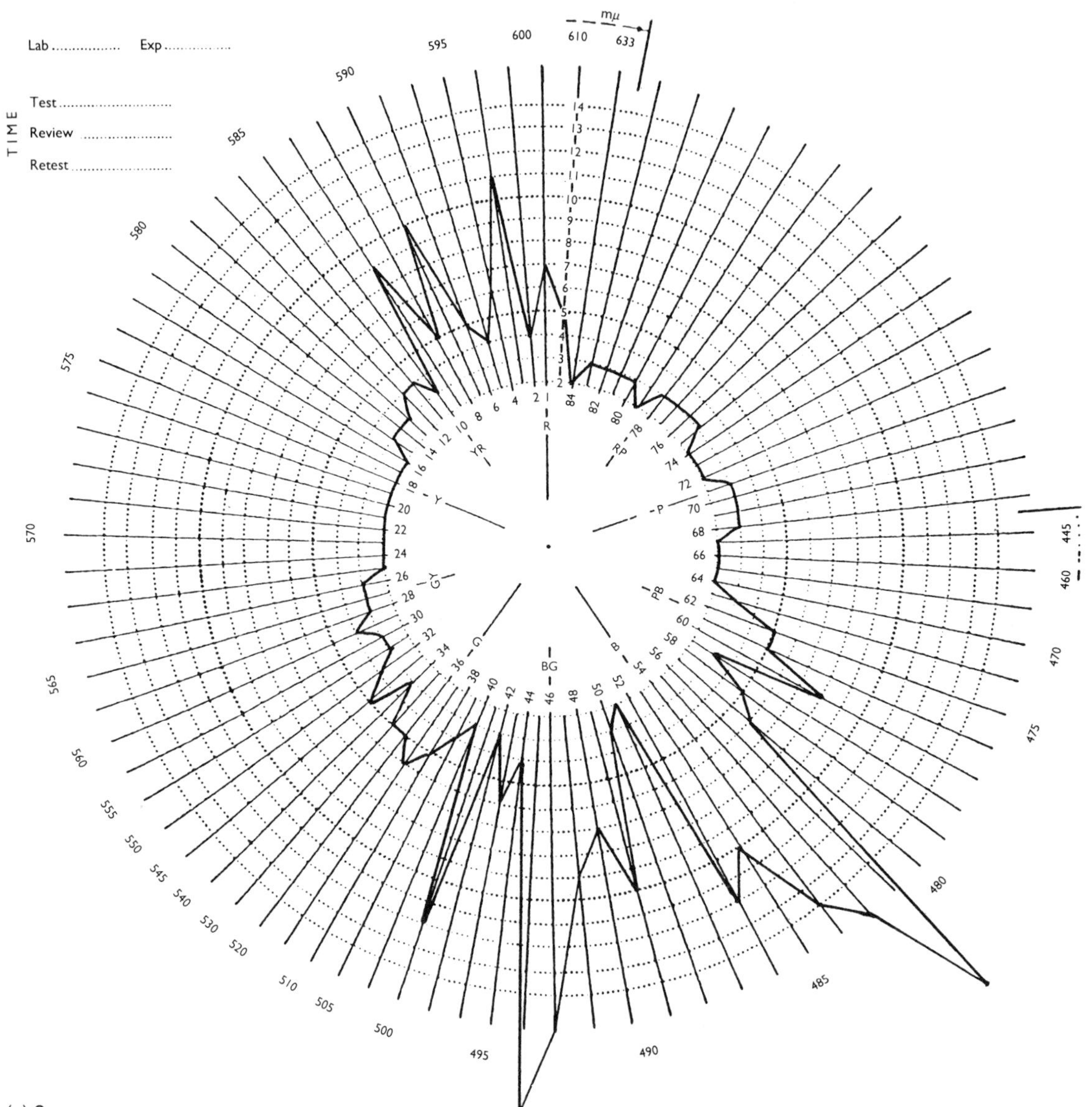

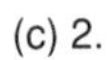
(c) 2.

Table 7.5 Position of the centre caps characterizing the axis of confusion for congenital dichromats on the Farnsworth–Munsell 100 hue test

	Centre cap	Range
Protanopes	17	(15–26)
	64	(58–68)
Deuteranopes	15	(12–17)
	58	(53–60)
Tritanopes	5	(4–6)
	45.5	(45–46)

From Dain and Birch 1987.

to the asymmetry of the colour spacing, typical rod monochromats tend to make more severe errors in the blue–green quadrant of the hue circle. However some monochromats are able to place the F–M 100 hue samples in a lightness scale and an apparent 'axis of confusion' is obtained which is intermediate between the tritan and deutan axes.

Farnsworth showed that individuals with normal colour vision can be divided into groups having superior, average, and poor hue discrimination according to the error score obtained with the test. Experienced colour matchers can obtain significantly better results and other people can improve their test performance after training or familiarity with the test. Error scores may improve by as much as 30 per cent on a second examination. The same individual variations in test performance are found with colour deficient observers and possible improvements in the error score, due to familiarity, have to be taken into account when repeated measurements are made. This is an important consideration when using the F–M 100 test to monitor acquired colour deficiency.

People with identical colour deficiency (such as dichromats) may have superior, average, or poor hue discrimination ability in the same way as normal observers. In these cases, differences in the error scores demonstrate the individual's ability to use careful observation to obtain good results. It is precisely this ability which the F–M 100 hue test is designed to assess. The distribution of error scores obtained by protanopes and deuteranopes is positively skewed and overlaps the distribution of scores obtained by anomalous trichromats of the same type (Fig. 7.6). Error scores obtained by anomalous trichromats show a weak correlation with the matching range found with the Nagel anomaloscope and the prominence of the axis of confusion is a better indicator of the severity of colour deficiency. An error score of more than 100 coupled with an axis of confusion shows significant colour deficiency. Approximately 50 per cent of anomalous trichromats do not obtain an axis of confusion and have an error score of less than 100. This result usually corresponds with a Nagel-matching range of less than 15 scale units (Fig. 7.7). Similar results are obtained by normal trichromats with poor hue discrimination ability.

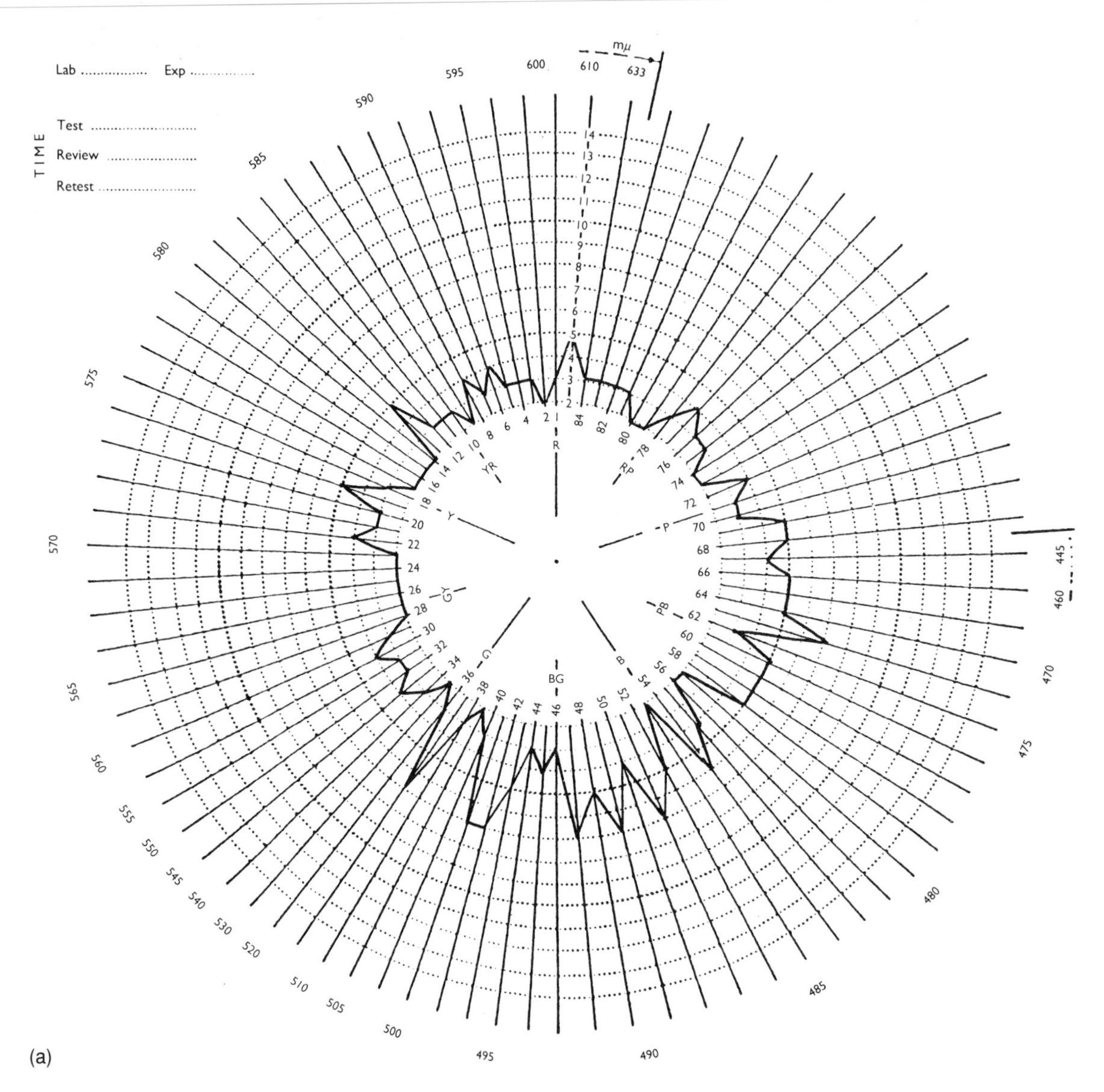

Fig. 7.5 Poor overall hue discrimination shown by the Farnsworth–Munsell 100 hue test. (a) Moderate overall reduction in hue discrimination (error score 136). (b) Severe overall reduction in hue discrimination (error score 320).

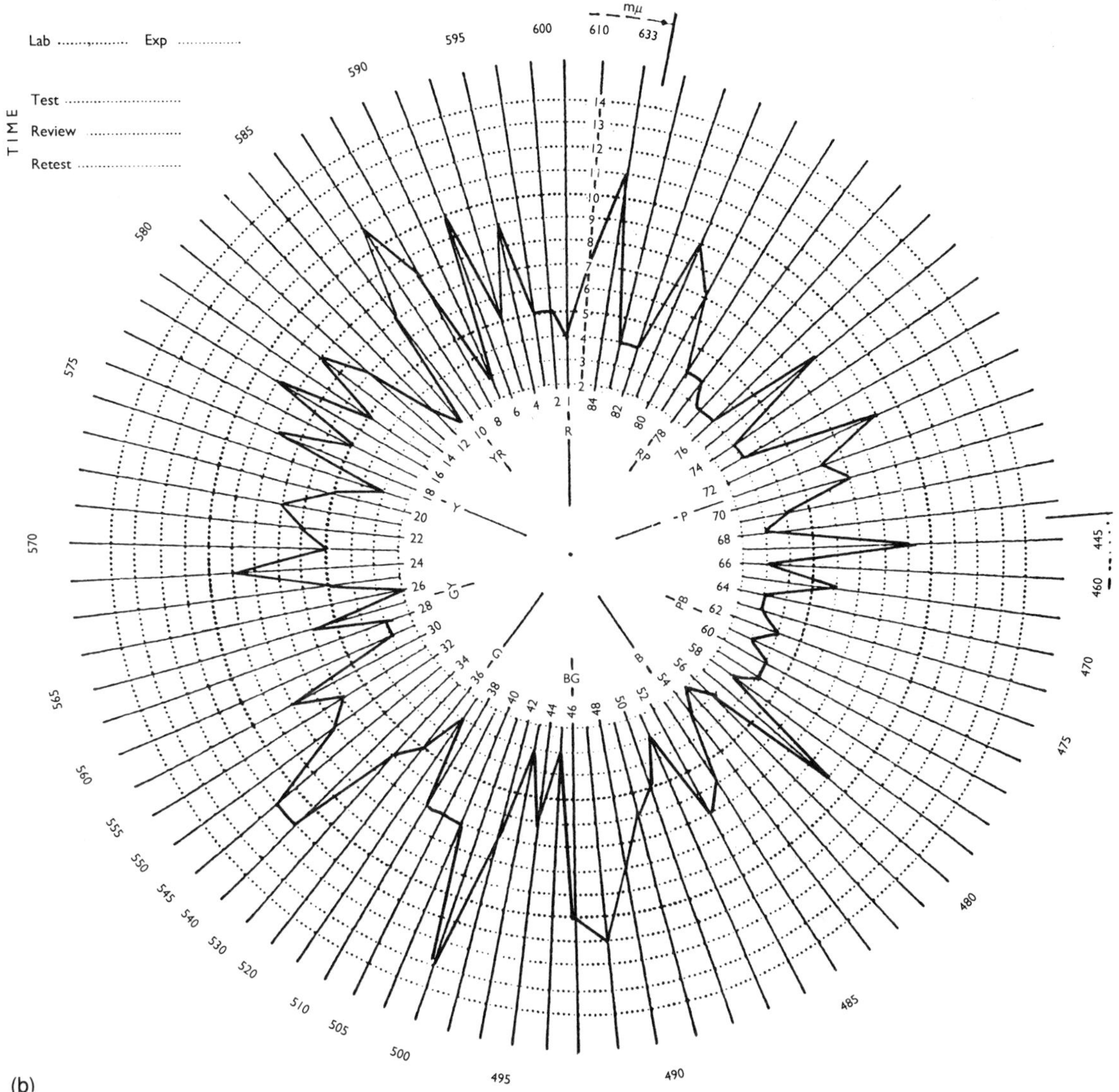

(b)

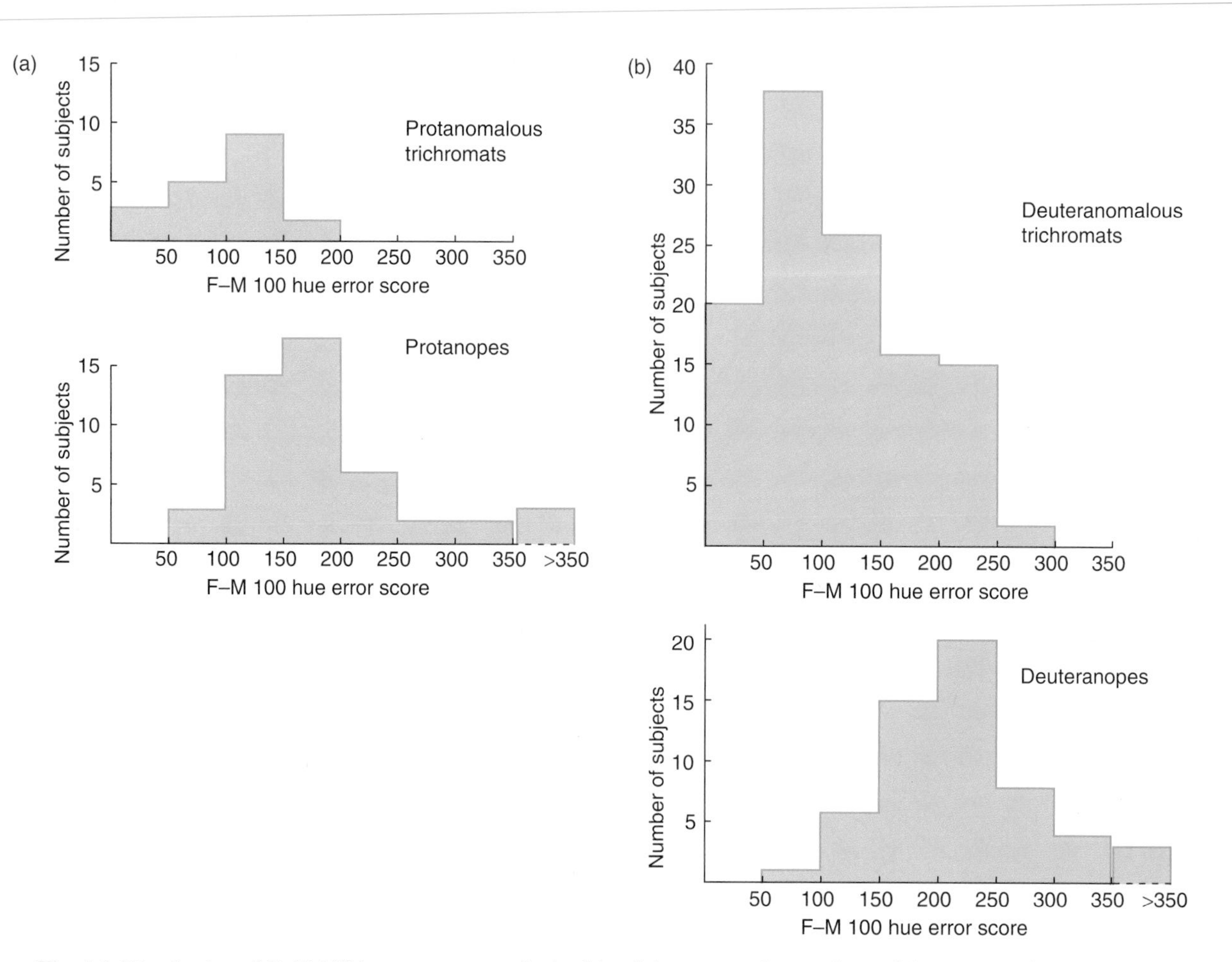

Fig. 7.6 Distribution of F–M 100 hue error scores obtained by dichromats and anomalous trichromats. (a) Error scores for 47 protanopes and 17 protanomalous trichromats. (b) Error scores for 57 deuteranopes and 117 deuteranomalous trichromats.

The F–M 100 hue error scores made by protan and deutan observers form two separate distributions. Error scores obtained by protans are lower than those found for deutans. This is because the relative luminous efficiency of protans is very different from that of normal trichromats and perceived luminance differences between the colours can be utilized to obtain a good result. Protans have more occupational problems than deutans because of the reduced visibility of red hazard warnings. It is therefore important to use a battery of colour vision tests to identify people with protan defects rather than rely exclusively on the F–M 100 hue error score as the basis for occupational advice.

Several methods have been proposed to maximize extraction of information from F–M 100 hue results. The scoring system invites statistical analysis and in recent years the test has become a tool to demonstrate analytical techniques. These techniques often disregard the clinical nature of the test and ignore the fact that the test/retest error scores may be significantly different. Clinically, the most useful methods of analysis extract information about a possible axis of confusion when the error

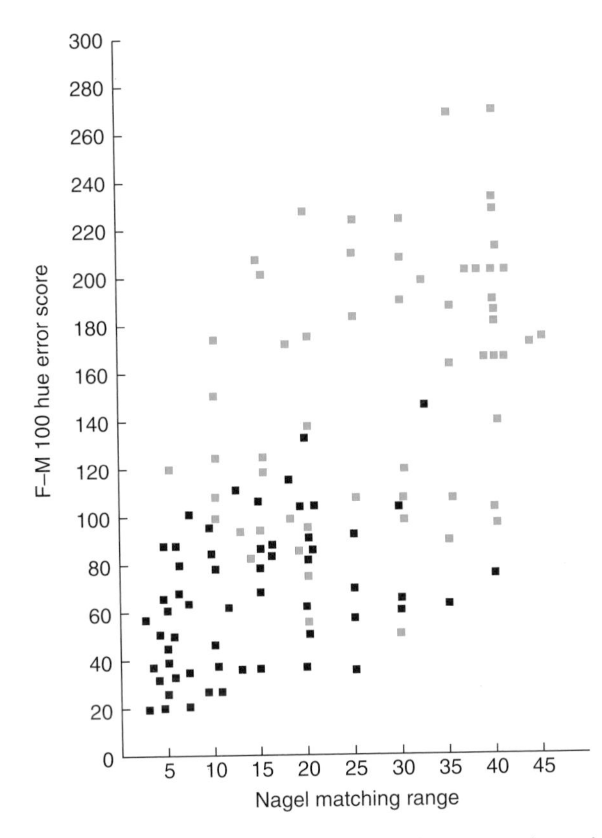

Fig. 7.7 F–M 100 hue error scores for deuteranomalous trichromats compared with the matching range of the Nagel anomaloscope. Error scores obtained by 117 deuteranomalous trichromats compared with the Nagel matching range in scale units. □ = Axis of confusion; ■ = no axis of confusion.

score is large and an axis is difficult to determine by visual inspection of the polar diagram. These involve averaging techniques which eliminate poor overall hue discrimination, or 'background noise', from the results so that any polarity is revealed. Averaging methods can be used to analyse individual results or group file data. Other techniques include Fourier analysis and vector analysis. Both these methods have disadvantages. For example, Fourier analysis imposes a cyclic variation on the results when the position of the axes of confusion are not symmetrical in the hue circle. Vector analysis provides information about the exact position of an axis of confusion. This emphasizes average dichromatic data and ignores inter-observer variations due to macular pigment density.

The need to provide population statistics has led to the almost indiscriminate use of the square root of the error score. This transformation has been found, empirically, to remove the 'skewness' inherent in group data although there are no grounds for supposing that the test results follow a square root law. Very large standard deviations from the mean are found when square rooting is carried out for groups of observers.

Even so, differences in square roots have been used to estimate the significance of changes in the error score in sequential measurements. These suggest that a change in error score is significant at the 0.05 level if the difference in the square roots exceeds 2.27 and at the 0.01 level if the difference exceeds 2.99 (Aspinall 1974). These figures are only a guide and confirmation of a change in colour vision should be sought from other grading tests. An alternative approach is to consider that a difference of more than 30 per cent in the error scores may be regarded as significant.

A test manual is provided for the F–M 100 hue test but the clinical use of the test is not fully explained.

The 28 hue test and the 40 hue test

Neither the 28 hue test nor the 40 hue test have been used extensively. Isochromatic colour confusions and poor discrimination of adjacent colours are possible and some people find that there are too many alternative arrangements for them to construct a natural colour order. Only one reference colour is provided which is intended to be both the first and last colour in the arrangement. Many observers find that they cannot 'join up' the colour order in this way and fail to complete a hue circle.

Sahlgren's saturation test

Sahlgren's saturation test is intended for the analysis of acquired colour deficiency but can be used to evaluate congenital red–green defects as well as tritan defects. The test has not been used extensively. There are 12 colour samples taken from the Swedish natural color system, which is similar to the Munsell system. Each colour is held in a cap which subtends 3.5° at a test distance of 30 cm. The samples are therefore considerably larger than those used in Farnsworth–Munsell tests and are intended to compensate for poor visual acuity in patients with acquired colour deficiency. There are 5 greenish–blue colours and 5 purplish–blue colours each representing a saturation scale, and 2 greys. The observer must sort the colours into groups which appear either greenish, purplish, or grey. Most people assume that the same number of samples should be allocated to each group and normal trichromats should be allowed to make 2 errors in selecting the most desaturated colour from both series as grey.

Anomaloscopes

The diagnostic value of a colour match between a mixture of red and green wavelengths and a spectral yellow was discovered by Lord Rayleigh in 1881. A spectral anomaloscope is used to distinguish normal and defective red–green vision and to diagnose the type of colour deficiency. The Nagel anomaloscope is the instrument of choice. Several other anomaloscopes have been produced as alternatives to the Nagel anomaloscope but some are much less accurate. A variety of different

Plate 1 The Holmgren Wool test.

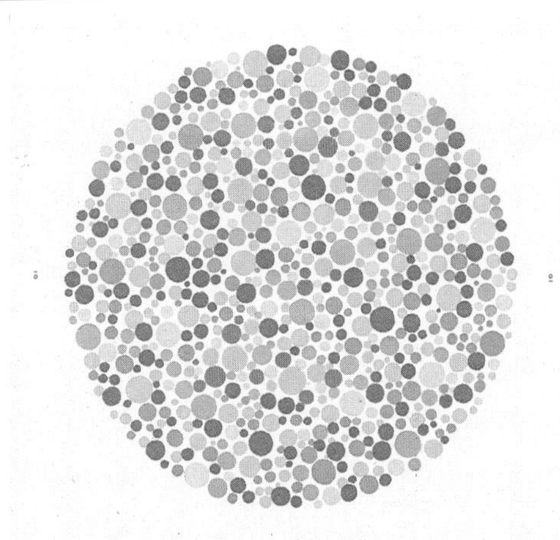

Screening plate with 'vanishing design'.

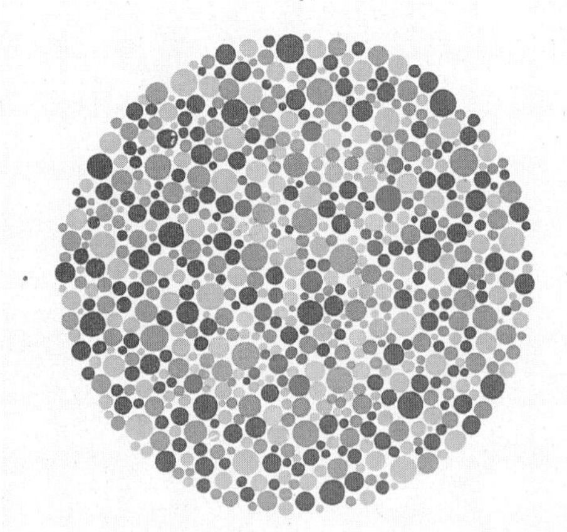

Screening plate with 'transformation design'.

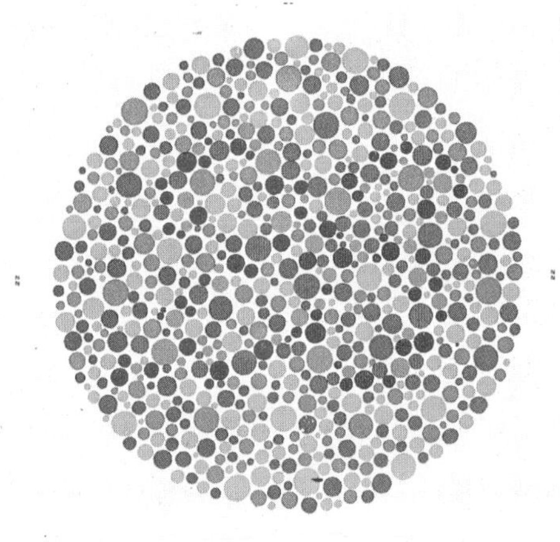

Protan/deutan classification plate.

Plate 2 The Ishihara test

(a) Grading.

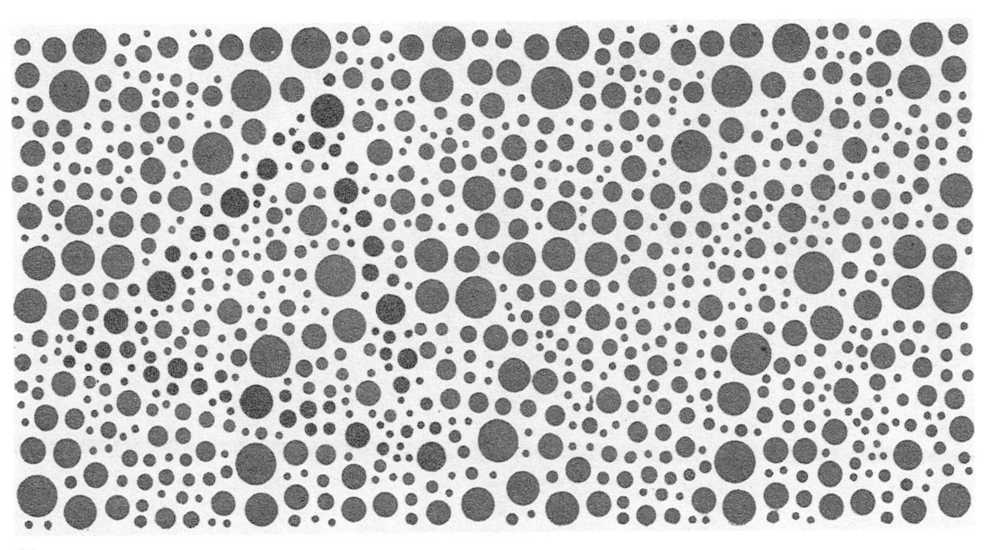

(b) Screening.

Plate 3 The new City University tritan test (vanishing design).

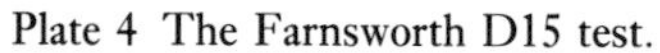

Plate 4 The Farnsworth D15 test.

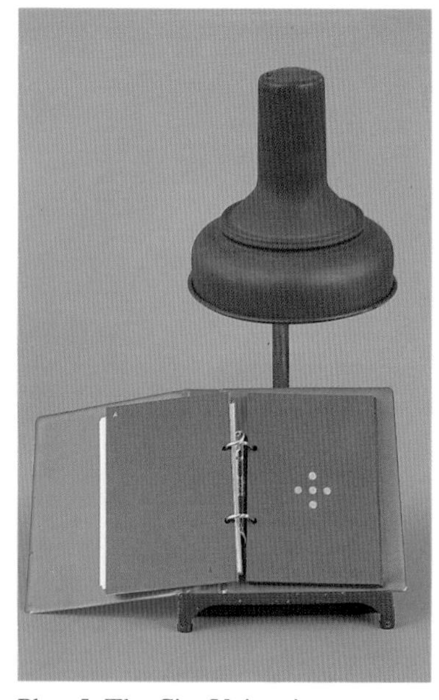

Plate 5 The City University test illuminated with the MacBeth Easel lamp.

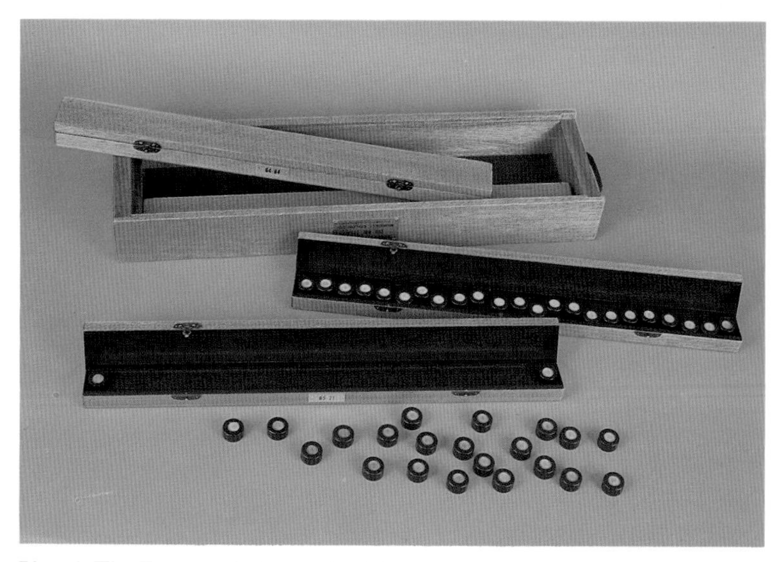

Plate 6 The Farnsworth–Munsell 100 hue test.

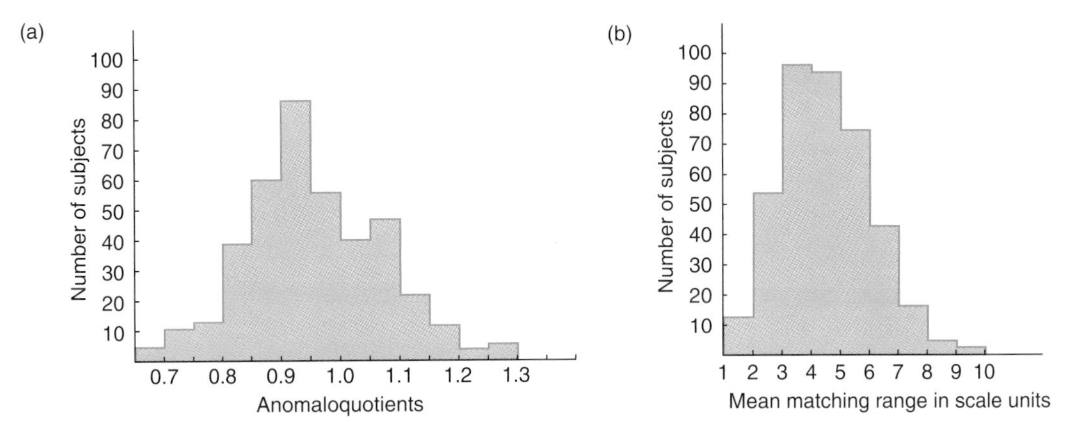

Fig. 7.8 Normal data for the Nagel anomaloscope. (a) Distribution of anomaloquotients derived from the mean matches of 400 normal trichromats. (b) The normal matching range in scale units derived from matches made by 400 normal trichromats.

methods for obtaining the colours and the colour mixture have been explored and some newer instruments provide a colour match intended to classify tritan defects. None of these instruments have been used as widely as the Nagel anomaloscope.

The Nagel anomaloscope

The Nagel anomaloscope was introduced in 1907. The instrument consists of a Maxwellian view spectroscope in which two halves of a 3° circular bipartite field are illuminated respectively by monochromatic yellow (589 nm) and a mixture of monochromatic red and green wavelengths (670 and 546 nm). A system of reciprocating slits keeps the luminance of the mixture field constant for any red–green ratio. The examination procedure is in two steps. In the first step, the observer makes several exact matches by adjusting both the red–green ratio and the luminance of the yellow test field. In the second step, the examiner sets the red–green ratio and the observer ascertains whether an exact match can be made by altering the luminance of the yellow test field. The matching range is recorded from the matching limits on the red/green mixture scale. Normal trichromats make a precise colour match within a small range of red–green ratios. Examination of a large number of normal observers establishes the mean and normal matching range for a particular instrument (Fig. 7.8). The matching ranges of protanomalous and deuteranomalous trichromats are outside the normal range of measurements and form two separate distributions (see Fig. 4.5, p. 39). Protanomalous trichromats require significantly more red light in their colour mixture and deuteranomalous trichromats more green. The extent of the matching range shows the severity of the defect and by inference the alteration in sensitivity of the affected photopigment. When a match is obtained, the yellow luminance setting provides information about relative luminous efficiency. Protans lower the luminance of the yellow test field when

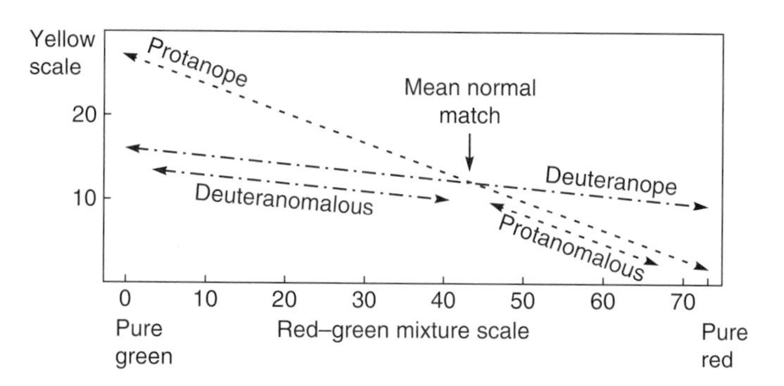

Fig. 7.9 Colour matches obtained with the Nagel anomaloscope in protan and deutan deficiency. Protanopes and deuteranopes are able to match all red/green ratios by adjusting the the luminance of the yellow comparison field. Protanomalous trichromats obtain matching ranges with an excess of red and deuteranomalous trichromats obtain matches with an excess of green in the matching field. The normal match is not accepted in anomalous trichromatism.

making a match with an excess of red showing that their relative luminous efficiency is reduced at long wavelengths. Protanopes and deuteranopes have only one photopigment in the spectral range provided by the instrument and are able to match any red–green ratio with yellow. The two types of dichromat are distinguished by the yellow luminance settings when matching pure red and pure green. Protanopes set a very low luminance for the yellow comparison field when matching pure red but deuteranopes match pure red and pure green with approximately the same luminance values (Fig. 7.9). Only extreme anomalous trichromats, who have abnormal long and medium-wave photopigments, obtain matches which encompass the normal match without including all red–green ratios.

An 'anomaloquotient' can be calculated from the midpoint of the matching range. This measurement was introduced to compare results obtained with different instruments which have identical wavelengths but numerically different normal matching ranges. Calculation of an anomaloquotient is also useful for comparing results obtained with the same instrument after the light source has been replaced. Anomaloquotients can be used to define the limits of the normal matching range (Fig. 7.8a), but a single colour match followed by calculation of an anomaloquotient is inadequate for classifying colour vision and an anomaloquotient is meaningless when the matching range is large. An anomaloquotient is calculated by dividing the numerical value of an individual mean match by the mean of a large group of normal observers. For example, if the mean normal match is at 42, on a red–green mixture scale which has pure green at 0 and pure red at 73, and that for an individual observer is at 40; the anomaloquotient for the observer is 33/40 divided by 31/42 or 1.12. People who have a greater proportion of green in their match, compared to the mean, have an anomaloquotient greater than unity and people with a greater proportion of red have an anomaloquotient less than unity. Ninety-five per cent of normal observers have a matching range of less than 6 scale units and anomaloquotient between 1.42 and 0.76.

A mark 2 Nagel anomaloscope was briefly available which provided an additional colour match to evaluate tritan defects. This was found to be ineffective and the instrument was withdrawn.

A test manual is provided which describes the design of the instrument but not it's clinical use.

The Neitz anomaloscope

The Neitz Anomaloscope is similar to the Nagel anomaloscope except that interference filters are used to provide the test and mixture wavelengths. Although the test field is darker than that of the Nagel anomaloscope, comparable results have been reported and the Neitz anomaloscope is considered to be a reliable substitute for the Nagel anomaloscope in clinical work.

The Besançon anomalometer

The Besançon anomalometer is a four-channel instrument which provides both a Rayleigh and a Moreland match. The colours are obtained with interference filters. The wavelengths chosen for the Rayleigh match are the same as in the Nagel anomaloscope. Earlier models of the instrument have slightly different wavelengths for the Moreland match but in the current instrument, the test field consists of a mixture of blue–green light (480 nm) and a desaturating yellow wavelength (580 nm). The matching wavelengths are 516 nm (green) and 470 nm (blue). These wavelengths have been selected to minimize the effects of individual variation in macular pigment density. The instrument has been used to study acquired colour deficiency in different types of ocular pathology. The Spectrum color vision meter 712 has been developed from the Besançon anomalometer. This instrument is interfaced with a computer and the examination routine is fully automated. The observer is presented with a series of mixture ratios and presses one of two buttons to record whether the mixture field matches the test field or not.

The Pickford–Nicolson anomaloscope

The Pickford–Nicolson anomaloscope is a direct vision instrument which employs broad-band filters to provide the test colours. Three colour matches can be made. These are a Rayleigh match, an Engelking–Trendelenberg blue–green match, and a Pickford–Lakowski match consisting of a mixture of blue and yellow to produce white. A variety of different field sizes and configurations are provided. The luminance of each colour is controlled by a separate shutter and the luminance of the test field does not remain constant as the ratio of the mixture colours is changed. The observer is positioned 1.5 m from the instrument and does not have access to the matching controls. The amount of ambient illumination affects the appearance of the test colours. The examiner adjusts the instrument controls according to the observers instructions until a match is obtained. The results are much less accurate than those obtained with instruments having spectral stimuli. The matching ranges of anomalous trichromats include the normal matching range and extend on both sides of the mixture scale. Reliance has to be placed on calculation of standard deviations and on luminance values at the limit of the matching range to classify protan and deutan observers. Dichromats are not readily distinguished from anomalous trichromats. The Pickford–Nicolson anomaloscope has been used to evaluate acquired colour deficiency and is no longer produced.

The A–N 59 anomaloscope (CIS) is similar in construction to the Pickford–Nicolson anomaloscope but was replaced in 1990 by a chromatest containing improved broad-band filters.

The Oscar test

The Oscar colour vision test has a unique design. The examination procedure involves the setting of minimum perceived flicker as the proportion of red and green light is changed. The observer views a small test field illuminated with a mixture of red (650 nm) and green (560 nm) light which is modulated in counter phase at about 16 Hz. The observer reduces the flicker to a minimum by varying the red–green modulation ratio. The instrument was designed as an alternative to an anomaloscope, but clinical trials have shown that deuteranomalous trichromats are not always identified and that dichromats cannot be distinguished from anomalous trichromats. The instrument is hand-held and the angular subtends of the test field cannot be accurately defined. The amount of ambient illumination present alters the appearance of the test field and influences the results obtained. The instrument is useful for classifying protans if no other means is available.

Other colour matching tests

The Davidson and Hemmendinger color rule

The Davidson and Hemmendinger color rule is derived from the Glenn color rule and is used by lighting engineers to evaluate the colour quality of lighting installations. The test consists of a slide rule with two painted colour panels. One panel is coloured in steps from pink through grey to blue and the other from green through grey to violet. The match obtained between the two panels defines the colour rendering properties of the illumination. The matching range obtained under standardized illumination, such as illuminant C, can be used to classify the colour vision of the observer.

The Sloan achromatopsia test

The Sloan achromatopsia test consists of seven cards each containing a series of grey rectangles representing a lightness scale. A coloured spot is positioned at the centre of each rectangle in six of the cards; the seventh panel contains grey spots and is used to introduce the test. The six colours are red, yellow–red, yellow, green, purple–blue, and red–purple. Monochromats can identify an exact match, between the spot and its surround, in each of the panels. The results correspond with the relative luminous efficiency of the observer's eye. The Sloan achromatopsia test is invaluable for the diagnosis of monochromatism and in distinguishing typical and atypical forms.

Adaptation tests for tritan defects

The TNO test and the Berkeley color threshold test

Two adaptation tests employing detection thresholds have been developed to isolate and measure the short-wave 'blue' mechanism in congenital tritan defects and acquired type 3 defects. Both the TNO test and the Berkeley color threshold test provide a large yellow adapting field, of high luminance, subtending between 10 and 15°. The yellow field adapts the medium and long-wave cone systems so that the remaining threshold wavelength sensitivity is mainly due to the short-wave 'blue' mechanism. Threshold sensitivity is measured for a single wavelength by varying the intensity of a superimposed blue light subtending approximately 1°. In the TNO test, the yellow background is provided by a 'cut-off' filter which only transmits wavelengths greater than 560 nm and an interference filter is utilized for the blue test wavelength of 460 nm. This instrument has been used in population studies as well as in the study of acquired type 3 defects found in diabetic retinopathy. The Berkeley color threshold test has been developed from a carousel slide projector and incorporates broad-band filters. Three alternative 'forced choice' presentations are made. The yellow upon yellow configuration is a test for red–green colour deficiency, and blue upon yellow is for tritan and type 3 defects. The additional presentation of yellow upon blue is used to assess pre-retinal absorption. The prototype instrument has been used to study acquired colour deficiency in diabetes and glaucoma.

Neither of these tests is commercially available but both can be individually constructed from published data (van Norren and Went 1981; Adams *et al.* 1987).

Lantern tests

Colour vision lanterns are vocational tests which simulate practical signal recognition under controlled viewing conditions. Lanterns were originally made to select railway workers but have since been adopted by the armed services and maritime transport systems. Lanterns are also used to examine air transport personnel and for licensing professional and recreational pilots. Most countries have developed lanterns to fulfil national colour vision requirements which are similar in design to the lanterns available in the UK (Cole and Vingrys 1982). There are two types; those which show single colours and those which show colours in pairs.

The Edridge–Green lantern (UK)

The Edridge–Green lantern was introduced in 1891 for selecting railway personnel. It has a very complex design and is now little used. The lantern consists of 5 rotating discs each containing 8 positions. Disc 1 contains different sized apertures and discs 2, 3, and 4 contain coloured filters. There are 8 different colours; two reds, two greens, white, yellow, blue, and purple. The intensity of the colours can be varied by superimposing the discs. Colours can also be 'mixed' by superimposing different filters. The fifth disc contains a clear aperture, 5 neutral density filters, a ribbed glass (to simulate rain), and a frosted glass (to simulate mist). Blue and purple filters are included although these colours are not used for track signals. In practice only a small number of the filters and additional devices need be used. The test manual does not give sufficient guidance on how to select, or present, the colours or how to interpret the results. The lantern should be operated in the same way as the Giles–Archer lantern. Neutral 4 is superimposed with a red glass to detect shortening of the red end of the spectrum.

The Giles–Archer lantern (UK)

The Giles–Archer lantern is for routine investigation of signal recognition. There are two lanterns. The standard lantern contains 6 colours; yellow (Y), white (W) (clear), blue–green (BG), green (G), red (R), and dark red (DR). The aviation model contains two additional filters; a light green (LG) and a signal yellow (SY). Both light green and signal yellow are frequently named incorrectly by normal trichromats and should not be included in the examination. The lantern is viewed at 6 m in a darkened room. Lengthy dark adaptation is not required. The examiner selects the colour order, decides on the number of colours presented, and controls the viewing time. At least 20 colours must be shown and some experience is necessary to produce an effective screening test. The order of the colours should be planned and a record sheet designed. This ensures that the same colour sequence is always presented. The difference between white and yellow is shown before the test begins and the permitted colour names are explained to the candidate. It is not advisable to show the colours in the order in which they are mounted on the filter wheel. White or yellow should be frequently interposed between the reds and greens in the sequence to produce successive colour contrast (Table 7.6). Each colour is shown for about 5 s and the filter aperture is covered while the next colour is selected. The test is made with the large aperture (5 mm subtending 2.9') and only repeated with a small aperture if the results are indecisive. A single error results in failure of the test. Errors are as follows:

1 Any misnaming of red.
2 Any misnaming of green.
3 Yellow called either red or green.
4 Inability to distinguish dark red at large aperture (called 'no light').

The following are not errors:

1 White called yellow.
2 Yellow called white.
3 Inability to distinguish dark red at small aperture.

Accurate colour vision screening is achieved if the lantern is used correctly and typical results are obtained in different types of congenital red–green

Table 7.6 Typical errors in colour naming with the Giles–Archer lantern

Colour sequence	G	W	DR	R	Y	SG	Y	R	W	G	Y	DR	W	SG	Y	R	G	Y	G	W
Protanope	(W)	W	–	R	Y	(W)	Y	R	W	(R)	Y	–	W	(W)	Y	R	(Y)	(R)	G	W
Deuteranope	(Y)	W	R	R	Y	G	(R)	R	W	(R)	(R)	R	W	(W)	(R)	R	(R)	(G)	(R)	W
Protanomalous trichomat	(Y)	(G)	R	R	Y	G	Y	R	W	(Y)	Y	R	W	G	Y	R	(Y)	(G)	(Y)	W
Deuteranomalous trichomat	(W)	W	R	R	Y	G	(R)	R	W	G	(G)	R	W	G	Y	R	(W)	Y	G	W

Errors are shown in brackets. G = green; R = red; Y = yellow; – = nothing seen; SG = signal green; DR = dark red; W = white.

deficiency (Table 7.6). Most errors are in naming green and signal green but there is no significant difference between the colour names given by protans and deutans. Both use 'white', 'yellow', and 'red'. Both protans and deutans misname yellow as either 'green' or 'red' depending on the preceding colour. The test demonstrates 'shortening of the red end of the spectrum' in protan defects. Protanopes, and most protanomalous trichromats, fail to see dark red and say that no light is being shown. One of the problems inherent in colour naming tests is that protans can easily deduce that if the light looks very dark, or cannot be seen, it must be described as 'red'.

The Holmes–Wright lantern (UK)

The Holmes–Wright Lantern (1974) has replaced the Board of Trade lantern (1913) and the Martin lantern (1939), and is similar in design to these earlier instruments. The type A Holmes–Wright lantern, which shows colours in pairs, is the standard lantern used in the armed services, the merchant navy, and the civil aviation authority. The Ishihara plates and the Holmes–Wright lantern are used together and colour perception is categorized according to the results obtained (Table 7.7). The required colour perception category varies according to the branch of the service for which the candidate is being considered. The lantern displays red, green, and white lights only. There are two different reds and two different greens. The chromaticity coordinates of these colours are within internationally agreed specifications for signal colours and represent the range of colours which may be observed in practice. The lantern contains nine pairs of colours and is viewed at 6 m. Before beginning the test, the examiner demonstrates the visual task by showing red, green, and white lights and naming them correctly. The first examination is made in normal room lighting (approximately 200 lx) with the colours at high luminance (200 microcandelas, μcd). The nine pairs of colours are shown three times making 27 presentations in all. Each colour pair is shown for about 5 s and the observer names the top colour first. The test is then repeated after 15 min dark adaptation with the colours at low luminance (20 μcd).

Table 7.7 Colour perception standards in the UK Armed forces

Type of colour vision	Classification	Test results
Exceptional	C.P.1	Pass Ishihara test Pass lantern test at low luminance following dark adaptation
Normal	C.P.2	Pass Ishihara test Pass lantern test at high luminance
Slight colour deficiency	C.P.3	Fail Ishihara test Pass lantern test at high luminance
Severe colour deficiency	C.P.4	Fail Ishihara test Fail lantern test at high luminance Correct colour naming of simple items of stationery

The Holmes–Wright lantern is an effective screening test for red–green colour deficiency. Colour deficient observers fail the test in both normal room illumination and after dark adaptation. Colour naming errors fall into three categories according to the frequency with which they are made.

1 Green called 'white' and white called 'green'.
2 White called 'red' and green called 'red'.
3 Red called 'white' and red called 'green'.

People with defective colour vision always misname green as 'white' and white as 'green' in both conditions of observation. This is the only category of error made by about 80 per cent of deuteranomalous trichromats, 50 per cent of protanopes and protanomalous trichromats, and 10 per cent of deuteranopes in normal photopic illumination.

All colour deficient observers make more mistakes, usually in more categories, following dark adaptation. However, people with slight colour deficiency do not generally misname red as 'green' or vice versa. The lantern has no dark red light to distinguish protan defects and in some cases protanopes respond in exactly the same way as deuteranomalous trichromats. The number of mistakes and the number of error categories are not therefore a reliable guide to the type and severity of colour deficiency. Some colour deficient people develop naming patterns, based on initial guesswork, and this may help them to achieve a good result. Since red lights look very dark to people with protan defects it is easy for them to guess that dark lights should be called red. Different patterns of false colour naming may occur on re-examination and the number of errors may vary by as much as 20 per cent.

The Farnsworth lantern (US)

The Farnsworth lantern (Falant) is the standard test used by the US armed services, the US coastguard, and FAA medical examiners. Although a similar procedure to that of the Holmes–Wright lantern is adopted, the test parameters differ in several respects. Selected isochromatic colours are shown instead of actual signal colours and the pass/fail criterion depends on an error score. A single showing of the test sequence is recommended for colour vision screening and the sequence is only shown three times if initial errors are made. An error score is derived from the average number of mistakes made on three sequences of the nine colour pairs. A score of 1.5 or more is considered as failure of the test. An error in naming either one or both colours in the pair is considered as one error. The test is performed in a normally illuminated room at a test distance of 8 ft, which is less than half the viewing distance used for the Holmes–Wright lantern. Different studies have shown that a substantial number of colour deficient people pass the test (between 15 and 50 per cent) and it appears that the pass rate has been set to give close agreement with failure of the Farnsworth D15 test. However there is no guarantee that dichromats will always fail the lantern test. Use of an error score does not take into account the category of naming error. For example, interchanging red and green colour names is likely to have more serious consequences than other mistake categories.

Clinical test batteries

Colour vision tests are designed to fulfil different functions and several tests may be needed to evaluate a colour vision defect. The tests selected depend on the priorities of the examination. These many include the need for screening, grading, classifying, and diagnosing the type of colour deficiency; distinguishing congenital and acquired colour deficiency, and making judgements about occupational suitability. Different combinations of tests are selected according to these priorities. The intellectual capabilites of the person being examined, the time available for the examination and the test location may also dictate the tests to be used. Availability and cost are additional limiting factors. A single test may be sufficient to identify red–green colour deficiency but a battery of tests is always required if occupational advice is needed and for the assessment of acquired colour deficiency.

Single tests for colour vision screening

A spectral anomaloscope, such as the Nagel anomaloscope, is a reliable screening test for protan and deutan colour deficiency and can be used to obtain a precise diagnosis of the type and severity of the defect. The Ishihara test is an efficient clinical screening test for red–green colour deficiency and there are several pseudoisochromatic test which have similar screening accuracy. Classification of protan and deutan defects is also obtained. The TNO test and the New City University tritan test are effective tritan screening tests.

Table 7.8 Severity of red–green colour deficiency shown by combined use of the Ishihara pseudoisochromatic plates and either the Farnsworth D15 test or the City University test

Ishihara plates	Farnsworth D15 test	The City University test	Severity of red–green colour deficiency
Fail	Pass	Pass	Slight
Fail	Fail with partial errors	Fail with 4 errors or less	Moderate
Fail	Fail with complete errors	Fail with 5 errors or more	Severe

Paired tests as a minimal test battery

Two tests are needed to identify red–green defects and tritan defects. Thus, the Ishihara test or the Nagel anomaloscope is paired with either the New City University tritan test or the TNO test.

Estimation of the severity of red–green deficiency is important for assessing practical colour ability. In this case the Ishihara plates and the Nagel anomaloscope can be used as a pair; only people who fail the plate test are examined with the more complicated anomaloscope procedure and the type and severity of colour deficiency is determined from the anomaloscope matching range. There are three clinical tests which can be paired with the Ishihara test to grade the severity of colour deficiency and to confirm the protan/deutan classification. These are the HRR plates, the Farnsworth D15 test and the City University test. All three tests detect moderate tritan defects. The HRR test is currently out of print and replacement copies cannot be obtained. The D15 test is an ideal choice for the second test but the format may be unsuitable for the test location or for the population being examined. The City University test is the third choice. The City University test is easy to use but has the disadvantage that the protan/deutan classification is uncertain. The appropriate use of the Ishihara test combined with either the D15 panel or the City University test is shown in Table 7.8.

The Ishihara plates and a colour vision lantern, such as the Holmes–Wright lantern, are frequently used together to select personnel for occupations requiring the correct identification of signal lights (Table 7.7). However, lanterns are not reliable grading tests and it is preferable to use additional tests to obtain information about the type and severity of colour deficiency.

A combination of the Ishihara plates and the F–M 100 hue test is often used to select people for occupations requiring normal colour vision and good colour aptitude, or to distinguish colour deficient people who have

Table 7.9 Selection of tests for a basic colour vision test battery

Test	Function
The Nagel anomaloscope	Classification of normal colour vision and diagnosis of protan and deutan deficiency. Dichromats are distinguished from anomalous trichomats
The Ishihara pseudoisochromatic plates	The most accurate clinical screening test for protan and deutan deficiency
The new City University tritan plates	A new pseudoisochromatic tritan screening test
The American Optical Company (HRR) pseudoisochromatic plates	Limited colour vision screening ability. Used as a grading test for protan, deutan, and tritan defects. Slight, moderate, and severe categories are distinguished
The Farnsworth D15 test and the City University test	A grading test. Individuals with slight colour deficiency pass, those with severe protan, deutan, or tritan deficiency fail
The Farnsworth–Munsell 100 hue test	A test of practical hue discrimination ability. Severe protan, deutan, and tritan defects are characterized by an axis of confusion

satisfactory hue discrimination ability for a particular job. In both cases a pass/fail level is based on an acceptable error score.

A test battery for a complete colour vision examination

A small test battery consisting of a screening test, (the Ishihara plates), a grading test (the HRR plates, the D15 test, or the City University test), and a diagnostic test (the Nagel anomaloscope) provides a complete analysis of congenital red–green deficiency (Table 7.9). The addition of the F–M 100 hue test provides information about practical hue discrimination ability and the Holmes–Wright lantern gives further information about signal recognition. If all these tests are included, the examination takes about an hour to complete.

A test battery suitable for the evaluation of acquired colour deficiency omits the lantern but includes all the other tests. In addition, pseudoisochromatic plates for the detection of tritan defects are required, such as the new City University tritan test or the SPP 2 plates, and a desaturated D15 test may be useful. A diagnostic test for tritan defects such as the TNO test or the Besançon anomaloscope is also desirable (Table 7.10). A degree of flexibility is needed for the examination of acquired colour deficiency and the test selection may be changed after

Table 7.10 Recommended test batteries for different functions

Function	Tests
(a) Screening for red–green colour deficiency	An appropriate illuminant The Ishihara plates (or the Dvorine plates) or the Nagel anomalosope
(b) Screening and grading red–green and tritan defects. Analysis of red–green defects	An appropriate illuminant The Ishihara plates The new City University tritan plates The AO (HRR) Plates (if available) The Farnsworth D15 test or the City University test The Nagel anomaloscope
(c) Evaluation of congential red–green defects and tritan defects	An appropriate illuminant The Ishihara plates The new City University tritan plates The AO (HRR) plates (if available) Saturated and desaturated D15 tests (or the City University test) The F–M 100 hue test The Nagel anomaloscope An adaptation test for tritan defects (A lantern test is included if required)
(d) Evaluation of acquired colour deficiency	An appropriate illuminant The Ishihara plates and the new City University plates or the SPP 2 plates The AO (HRR) plates (if available) Saturated and desaturated D15 tests The F–M 100 hue test The new color test The Nagel anomaloscope The Besançon anomaloscope or an adaptation test for tritan defects Measurement of visual acuity and visual fields

preliminary results have been obtained so that the investigation can proceed along specific lines.

Typical results for different types of congenital colour deficiency are shown in the examples which follow this chapter.

References

Adams, A.J., Schefrin, B.E., and Huie, K. (1987). A new clinical threshold test for eye disease. *American Journal of Optometry and Physiological Optics*, **64**, 29–37.

Aspinall, P.A. (1974). Inter-eye comparison and the 100 hue test. *Acta Ophthalmologica*, **52**, 307–16.

Belcher, S.J., Greenshields, K.W., and Wright, W.D. (1958). A colour vision survey. *British Journal of Ophthalmology*, **42**, 355–9.

Berson, E.L., Snadberg, M.A., Rosner, A., and Sullivan, P.L. (1983). Colour plates to help identify patients with blue cone monochromatism. *American Journal of Ophthalmology*, **95**, 6, 741–7.

Bowman, K.J. (1982). A method of qualitative scoring of the Farnsworth D15 panel. *Acta Ophthalmologica*, **60**, 907–16.

Cole, B.L. and Vingrys, A.J. (1982). A survey and evaluation of lantern tests of colour vision. *American Journal of Optometry and Visual Science*, **59**, 4, 346–74.

Dain, S.J. and Birch, J. (1987). An averaging method for the interpretation of the Farnsworth–Munsell 100 hue test. *British Journal of Physiological Optics*, **7**, 3, 267–80.

Farnsworth, D. (1943). The Farnsworth–Munsell 100 hue and dichotomous tests for colour vision. *Journal of the Optical Society of America*, **33**, 6, 568–78.

Favre, A. (1873). Du daltonisme du point de vue de l'industrie des chemins de fer. *Lyon Medicale*, **14**, 6–20.

Frey, R.G. (1962). Die Trennnscharfe einiger pseudoisochromatischer Tafelproben. *von Graefes Archives für Ophthalmology*, **165**, 20–30.

Edgridge-Green, F.W. (1891). *Colour blindness and colour perception*. Kegan, Paul, Trench and Tubner, London.

Murray, E. (1943). Evolution of colour vision tests. *Journal of the Optical Society of America*, **33**, 6, 316–34.

van Norren, D. and Went, L.N. (1981). A new test for the detection of tritan defects evaluated in two surveys. *Vision Research*, **21**, 1303–6.

Paulson, H. (1973). *Comparison of color vision tests used by the armed forces*., pp. 34–64 Academy of Science, Washington DC.

Seebeck, A. (1837). Über den bei mancher personner vorkcommenden Mangel or Farbesinn. *Annales von Physiology* (Leipzig), **42**, 177–233.

Taylor, W.O.G. (1975). Constructing your own P.I.C test. *British Journal of Physiological Optics*, **30**, 22–4.

Vingrys, A.J. and King-Smith, P.E. (1988). A quantitative scoring technique for panel tests of color vision. *Investigative Ophthalmology and Visual Science*, **29**, 50–63.

Wilson. G. (1853). Railroad signals and colour blind drivers and signalmen. Letter to the Atharaeum, London. April 2nd.

Further Reading

Boltz, C.L. (1952). *A statue to Mr Trattles and other scientific topics*. Butterworths, London.

Typical results obtained by people with different types of colour deficiency on a battery of colour vision tests and reports indicating occupational ability

Case example—protanope

Ref:		MGB
Sex:		Male
Age:		25
V/A:		6/5
Occupation:		Office Manager
Subjective comments:		Would like to join airline as cabin staff. Difficulty selecting matching clothes. Blue and pink papers difficult to distinguish. Classified as colour deficient at school.
Family history:		Maternal grandfather colour deficient.
		Results
Ishihara plates:		Fail. Protan defect.
AO, HRR plates:		Fail. Severe protan defect.
Standard D15:		Fail. Protan. (Fail—H16 test.)
D15 5/2:		–
TCU Test:		Fail. 2 protan errors. 2 deutan errors.
F–M 100 hue test:		Error score—152. Protan axis of confusion.
Nagel anomaloscope: (normal match 44±2)		Full matching range. PROTANOPE.
Holmes–Wright lantern:	(a)	PHOTOPIC. White called 'green'. Green called 'white'. White called 'red'.
	(b)	SCOTOPIC. Green called 'white'. White called 'green'. White called 'red'. Red called 'green'.
Comments:		Unlikely to be accepted for airline cabin staff as normal colour vision is necessary to operate safety equipment. (The TCU test gives a mixed protan/deutan classification as an artefact of the test design.)

Colour vision report

The results of the colour vision examination show that MGB has a colour vision defect known as protanopia. The defect is severe. A guide to practical colour vision capabilities is indicated by 'x'.

Colour recognition

1. Normal. The defect is of no practical significance. [ ]
2. Slightly impaired. Errors may occur with very pale colours or very dark colours. [ ]
3. Moderately impaired. Errors may occur with bright colours in some viewing conditions. [ ]
4. Poor. Gross errors occur in all viewing conditions. [x]

Sensitivity to red light

Normal [ ] Defective [x]

Hue discrimination

The ability to distinguish small colour differences is

Average [ ] Below average [ ] Poor [x]

General recommendations for occupations involving colour judgements.

1. Suitable for all types of work involving colour recognition except colour matching and colour quality control. [ ]
2. Suitable for most tasks involving colour recognition under good illumination and where speed is not important. [ ]
3. Suitable for tasks involving recognition of large colour differences only. [ ]
4. Not suitable for tasks requiring colour judgements. [x]

Case example—protanomalous trichromat

Ref:	MB
Sex:	Male
Age:	21
V/A:	6/4
Occupation:	Economics graduate seeking career.
Subjective comments:	Classified colour deficient at school. Has problems distinguishing traffic lights on a bright background and in seeing stop signals at night. Cannot distinguish black and red print. Has difficulty using the London Underground map as some colours look similar.
Family history:	Maternal grandfather colour deficient.
	Results
Ishihara plates:	Fail. Protan defect.
AO, HRR plates:	Fail. Moderate protan defect.
Standard D15:	Fail. Protan. (Pass—H16 test.)
D15 5/2:	–
TCU test:	Fail. 1 protan, 2 deutan, and 1 tritan error.
F–M 100 hue test:	Error score—132. Protan axis of confusion.
Nagel anomaloscope: (normal match 44±2)	Matching range 40–60. PROTANOMALOUS TRICHROMAT.
Holmes–Wright lantern: (a)	PHOTOPIC. White called 'green'. Green called 'white'.
(b)	SCOTOPIC. Green called 'white'. White called 'green'. Red called 'white'.
Comments:	Unlikely to be accepted as a management trainee by British Rail because normal colour vision is required while work shadowing engineers and drivers. (The TCU test gives a mixed classification as an artefact of the test design.)

Colour vision report

The results of the colour vision examination show that MB has a colour vision defect known as protanomalous trichromatism. The defect is moderate. A guide to practical colour vision capabilities is indicated by 'x'.

Colour recognition

1. Normal. The defect is of no practical significance. [ ]
2. Slightly impaired. Errors may occur with very pale colours or very dark colours. [ ]
3. Moderately impaired. Errors may occur with bright colours in some viewing conditions. [x]
4. Poor. Gross errors occur in all viewing conditions. [ ]

Sensitivity to red light

Normal [ ] Defective [x]

Hue discrimination

The ability to distinguish small colour differences is

Average [ ] Below average [ ] Poor [x]

General recommendations for occupations involving colour judgements.

1. Suitable for all types of work involving colour recognition except colour matching and colour quality control. [ ]
2. Suitable for most tasks involving colour recognition under good illumination and where speed is not important. [ ]
3. Suitable for tasks involving recognition of large colour differences only. [x]
4. Not suitable for tasks requiring colour judgements. [ ]

Case example—deuteranope

Ref:		RCD
Sex:		Male
Age:		40
V/A:		6/6
Occupation:		Lorry driver.
Subjective comments:		Not aware of colour vision problems. Never examined previously. Would like to join the police force.
Family history:		None known.
		Results
Ishihara plates:		Fail. Deutan defect.
AO, HRR plates:		Fail. Severe deutan defect.
Standard D15:		Fail. Deutan. (Fail—H16 test.)
D15 5/2:		–
TCU test:		Fail. 8 Deutan errors.
F–M 100 hue test:		Error score—200. Deutan axis of confusion.
Nagel anomaloscope: (normal match 44±2)		Full matching range. DEUTERANOPE.
Holmes–Wright lantern:	(a)	PHOTOPIC. White called 'green'. Green called 'red'. White called 'red'. Red called 'green'.
	(b)	SCOTOPIC. White called 'green'. Green called 'white'. White called 'red'. Red called 'green'. Green called 'red'.
Comments:		Will not be able to join the police force as normal colour vision is required. It is unusual for a deuteranope to be unaware of his colour deficiency even if he has not been examined before.

Colour vision report

The results of the colour vision examination show that RCD has a colour vision defect known as deuteranopia. The defect is severe. A guide to practical colour vision capabilities is indicated by 'x'.

Colour recognition

1. Normal. The defect is of no practical significance. [ ]
2. Slightly impaired. Errors may occur with very pale colours or very dark colours. [ ]
3. Moderately impaired. Errors may occur with bright colours in some viewing conditions. [ ]
4. Poor. Gross errors occur in all viewing conditions. [x]

Sensitivity to red light

Normal [x] Defective [ ]

Hue discrimination

The ability to distinguish small colour differences is

Average [ ] Below average [ ] Poor [x]

General recommendations for occupations involving colour judgements.

1. Suitable for all types of work involving colour recognition except colour matching and colour quality control. [ ]
2. Suitable for most tasks involving colour recognition under good illumination and where speed is not important. [ ]
3. Suitable for tasks involving recognition of large colour differences only. [ ]
4. Not suitable for tasks requiring colour judgements. [x]

Case example—deuteranomalous trichromat

Ref:		AT
Sex:		Male
Age:		44
V/A:		6/4
Occupation:		Chemistry lecturer.
Subjective comments:		Found colour deficient at school. Advised not to pursue a career in chemistry but was accepted at university. Had lots of problems distinguishing chemical reactions denoted by a colour change, now always uses instrumental methods for colour measurement. Seeks help for electrical wiring tasks.
Family history:		Brother colour deficient.
		Results
Ishihara plates:		Fail. Deutan defect.
AO, HRR plates:		Fail. Moderate deutan defect.
Standard D15:		Pass.
D15 5/2:		Fail. Deutan.
TCU test:		Fail. 4 deutan errors.
F–M 100 hue test:		Error score—100. Deutan axis of confusion.
Nagel anomaloscope: (normal match 44±2)		Matching range 0–40. DEUTERANOMALOUS TRICHROMAT.
Holmes–Wright lantern:	(a)	PHOTOPIC. Green called 'white'. White called 'green'.
	(b)	SCOTOPIC. Green called 'white'. White called green'. Red called 'white'.
Comments:		AT successfully completed his training as a chemist but encountered colour recognition problems. He therefore opted for an academic rather than an industrial career.

Colour vision report

The results of the colour vision examination show that AT has a colour vision defect known as deuteranomalous trichromatism. The defect is moderate. A guide to practical colour vision capabilities is indicated by 'x'.

Colour recognition

1. Normal. The defect is of no practical significance. [ ]
2. Slightly impaired. Errors may occur with very pale colours or very dark colours. [ ]
3. Moderately impaired. Errors may occur with bright colours in some viewing conditions. [x]
4. Poor. Gross errors occur in all viewing conditions. [ ]

Sensitivity to red light

Normal [x] Defective [ ]

Hue discrimination

The ability to distinguish small colour differences is

Average [ ] Below average [x] Poor [ ]

General recommendations for occupations involving colour judgements.

1. Suitable for all types of work involving colour recognition except colour matching and colour quality control. [ ]
2. Suitable for most tasks involving colour recognition under good illumination and where speed is not important. [ ]
3. Suitable for tasks involving recognition of large colour differences only. [x]
4. Not suitable for tasks requiring colour judgements. [ ]

Case example—slight deuteranomalous trichromat

Ref:	MAB
Sex:	Male
Age:	14
V/A:	6/4
Occupation:	At school.
Subjective comments:	No problems with colours. Failed school screening test and would like careers advice.
Family history:	Father colour deficient.
	Results
Ishihara plates:	Fail. Deutan defect. 7 errors only, 5 on confusion designs and 2 vanishing plates. Classification of deutan obtained from brightness difference of classification figures.
AO, HRR plates:	Borderline fail, 2 red–green screening errors only.
Standard D15:	Pass.
D15 5/2:	Pass.
TCU test:	Pass.
F–M 100 hue test:	Error score—100. Poor hue discrimination, possible slight deutan axis of confusion.
Nagel anomaloscope: (normal match 44±2)	Matching range 12–20. DEUTERANOMALOUS TRICHROMAT.
Holmes–Wright lantern:	(a) PHOTOPIC. Green called 'white'.
	(b) SCOTOPIC. Green called 'white'. White called 'green'.
Comments:	MAB fails the Ishihara plates but does not make all the possible errors. Classification of a deutan defect is made by brightness difference only. The Nagel anomaloscope clearly indicates a deuteranomalous defect. The other tests are either passed or give a borderline result. Advised of careers in which normal colour vision is mandatory. The colour deficiency of MAB's father is coincidental.

Colour vision report

The results of the colour vision examination show that MAB has a colour vision defect known as deuteranomalous trichromatism. The defect is slight. A guide to practical colour vision capabilities is indicated by 'x'.

Colour recognition

1. Normal. The defect is of no practical significance. [ ]
2. Slightly impaired. Errors may occur with very pale colours or very dark colours. [x]
3. Moderately impaired. Errors may occur with bright colours in some viewing conditions. [ ]
4. Poor. Gross errors occur in all viewing conditions. [ ]

Sensitivity to red light

Normal [x] Defective [ ]

Hue discrimination

The ability to distinguish small colour differences is

Average [ ] Below average [x] Poor [ ]

General recommendations for occupations involving colour judgements.

1. Suitable for all types of work involving colour recognition except colour matching and colour quality control. [ ]
2. Suitable for most tasks involving colour recognition under good illumination and where speed is not important. [x]
3. Suitable for tasks involving recognition of large colour differences only. [ ]
4. Not suitable for tasks requiring colour judgements. [ ]

Case example—slight deuteranomalous trichromat

Ref:		CW
Sex:		Male
Age:		24
V/A:		6/4
Occupation:		Medical practitioner.
Subjective comments:		Thinks that he is making mistakes with colour coded methods of glycosuria analysis. Previously passed colour vision screening tests. No general health problems or medical history of ocular or intracranial pathology.
Family history:		None known.
		Results
Ishihara plates:		Fail. Slow responses. 4 errors on confusion designs.
AO, HRR plates:		Borderline fail, 2 red–green screening errors only.
Standard D15:		Pass.
D15 5/2:		Pass.
TCU test:		Pass.
F–M 100 hue test:		Error score—16. No axis of confusion.
Nagel anomaloscope: (normal match 44±2)		Matching range 20–22. DEUTERANOMALOUS TRICHROMAT.
Holmes–Wright lantern:	(a)	PHOTOPIC. Green called 'white'.
	(b)	SCOTOPIC. Green called 'white'.
Comments:		CW has minimal deuteranomalous trichromatism. Both eyes have identical colour deficiency so it must be assumed that he was 'missed' at a previous colour vision examination and that the defect is congenital. Advised to take extra care with colour coded tasks and to seek help if in doubt.

Colour vision report

The results of the colour vision examination show that CW has a colour vision defect known as deuteranomalous trichromatism. The defect is slight. A guide to practical colour vision capabilities is indicated by 'x'.

Colour recognition

1. Normal. The defect is of no practical significance. [x]
2. Slightly impaired. Errors may occur with very pale colours or very dark colours. [ ]
3. Moderately impaired. Errors may occur with bright colours in some viewing conditions. [ ]
4. Poor. Gross errors occur in all viewing conditions. [ ]

Sensitivity to red light

Normal [x] Defective [ ]

Hue discrimination

The ability to distinguish small colour differences is

Average [x] Below average [ ] Poor [ ]

General recommendations for occupations involving colour judgements.

1. Suitable for all types of work involving colour recognition except colour matching and colour quality control. [x]
2. Suitable for most tasks involving colour recognition under good illumination and where speed is not important. [ ]
3. Suitable for tasks involving recognition of large colour differences only. [ ]
4. Not suitable for tasks requiring colour judgements. [ ]

Case example—tritanope

Ref:		JF
Sex:		Male
Age:		31
V/A:		6/5
Occupation:		Engineer.
Subjective comments:		Always aware of colour deficiency but passes screening tests and is not believed.
Family history:		None known. Three young children, one of whom makes mistakes naming colours.
		Results
Ishihara plates:		Pass.
AO, HRR plates:		Fail. Severe tritan defect.
Standard D15:		Fail. Tritan.
D15 5/2:		–
TCU test:		Fail. 8 tritan errors.
F–M 100 hue test:		Error score—140. Tritan axis of confusion.
Nagel anomaloscope: (normal match 44±2)		Normal red–green vision.
Holmes–Wright lantern:	(a)	PHOTOPIC. Pass.
	(b)	SCOTOPIC. Pass.
		TNO threshold at 460 nm: 1.6 log units.
		Wavelength matching 450 nm matches 502 nm.
		TRITANOPE.
Comments:		JF is a congenital tritanope.

Colour vision report

The results of the colour vision examination show that JF has a colour vision defect known as tritanopia. The defect is severe. A guide to practical colour vision capabilities is indicated by 'x'.

Colour recognition

1. Normal. The defect is of no practical significance. [ ]
2. Slightly impaired. Errors may occur with very pale colours or very dark colours. [ ]
3. Moderately impaired. Errors may occur with bright colours in some viewing conditions. [ ]
4. Poor. Gross errors occur in all viewing conditions. [x]

Sensitivity to red light

Normal [x] Defective [ ]

Hue discrimination

The ability to distinguish small colour differences is

Average [ ] Below average [ ] Poor [x]

General recommendations for occupations involving colour judgements.

1. Suitable for all types of work involving colour recognition except colour matching and colour quality control. [ ]
2. Suitable for most tasks involving colour recognition under good illumination and where speed is not important. [ ]
3. Suitable for tasks involving recognition of large colour differences only. [ ]
4. Not suitable for tasks requiring colour judgements but able to distinguish red–yellow–green colour codes. [x]

8. Examining children for colour deficiency

Colour deficiency is found in families and parents are alerted to the possibility that a child may have defective colour vision if another family member is affected. The most likely form of transmission, in an X-linked inheritance, is from maternal grandfather to grandson. Boys who have a colour deficient maternal grandfather, or uncle, are at risk, and boys who have a colour deficient brother have a 50 per cent chance of being similarly affected. Girls are not likely to be colour deficient unless there is a history of colour deficiency in both parent's families. Children with severe colour deficiency often make mistakes with colours at an early age. The child may use the wrong colour names for familiar objects or choose inappropriate colouring materials. Children with slight colour deficiency are unlikely to make obvious mistakes and, if there is no close relative affected, the first indication is from failure of a screening test.

Many occupations require good colour vision and careers advisers need to know about colour deficiency before discussing career options. However, if colour vision screening is delayed until adolescence the child may have been placed at an educational disadvantage earlier on and the sudden diagnosis of colour deficiency can be traumatic. Colour is an important coding dimension and is frequently employed as a teaching aid.

The earliest years of formal education are the most colourful. Colour creates an aesthetically pleasing classroom environment and stimulates awareness. Children are taught colour names and colour identification is used to learn new words and correct spelling. Children should be able to use a variety of colour names correctly by 5–6 years of age and colours of varying hue, saturation, and chroma can be matched with almost 100 per cent accuracy by 6 year olds. Facility with colours is therefore assumed in teaching materials intended for these age groups. Colour codes are often incorporated in reading programmes and it may be necessary to understand the principles of colour mixing in order to benefit fully, for example the child may need to know that mixing red and yellow pigments produces orange, and that blue and red make purple. Reading books of a certain standard are sometimes identified by a coloured band and colour coding may be used in simple science experiments. Red, yellow, green codes are the most popular and severely colour deficient children must be able to utilize luminance contrast differences to distinguish them. Colour is used to identify shapes and building blocks in number games. Units of the same colour are assembled to make sets and colour mixing laws may be invoked to demonstrate that the addition of two sets equals a third. Colours are often used to teach arithmetic and ten colours are needed for decimal systems. When there are as many as ten colours it is essential that luminance contrast differences can be used to aid identification. With all colour coded materials the extent of practical learning difficulties depends very much on the emphasis placed on colour identification by the teacher. Most teaching aids contain only three or four bright colours and only children with severe colour deficiency are likely to have problems with them. However some materials require green/brown discrimination which all colour deficient people find difficult. Teachers need to be aware of colour deficient pupils so that teaching methods can be adapted to minimize difficulties and to avoid misunderstandings.

In the UK, colour is used less as a teaching aid after 7 years of age but colour codes appear in specific areas of the curriculum as the child gets older. Colour recognition is important in understanding geographical maps and is essential for recognizing chemical reactions. Art, needlework, metalwork, and woodwork all involve colour. Engineering diagrams and biological specimens require colour recognition. Colour deficient pupils may need to ask for help in practical examinations in these subjects. Colour is used to emphasize and link different sets of information in a variety of ways in many different subject areas. Coloured histograms are a popular means of showing economic data and population statistics as well as the results of scientific experiments. Colour TV monitors offer another coding dimension.

The widespread use of colour in schools suggests the need for colour vision screening at the beginning of formal education and for information about colour vision problems to be included in records that are passed from one educational institution to another. Teachers and parents should be aware of possible difficulties and occupational limitations should be discussed well in advance of making a choice of career. It is important that the situation is explained tactfully. A child should not be made to feel that he or she is 'defective' and should not be exposed to ridicule from peers. Poor handling of the situation can lead to psychological trauma, resentment, or denial.

Individual colour vision examinations are made by optometrists and family doctors but large numbers of children are screened as part of a routine health check in schools. The age at which health screening is undertaken in the UK varies locally but periodic checks at about 5, 11, and 15 years of age are recommended. Colour vision screening is usually included at 11 years of age. Indeed a survey undertaken in 1976 showed that 70 per cent of Area Health Authorities wait until children are over 10 years of age before screening for colour deficiency and that the examination is mainly carried out by the school nurse (Voke 1976) (Table 8.1). Ideally colour vision screening should be included in the first medical examination

Table 8.1 Colour vision screening in UK schools (1976)

Age-group in years	Percentage of Local Authorities conducting first screening test
4–6	14%
7–9	14%
10–12	64%
13–15	8%

soon after formal education begins to correspond with the introduction of colour coded teaching materials. Alternatively, teachers could perform the examination themselves. Pseudoisochromatic plates are readily understood by young children and reliable results can be obtained by experienced examiners with children as young as 3 years of age. Screening 5-year-olds does not present too much difficulty for relatively inexperienced examiners and results are easily obtained with 7-year-olds.

All colour vision tests require comprehension of the visual task and different tests become useful at different stages of development. Chronological age is less important than the attainment of the necessary cognitive skills. However, screening tests designed specially for children have been largely unsuccessful because there has been too much emphasis on simplifying the visual task to make it easier for children to comprehend and insufficient attention paid to the accuracy of the colour design. With the exception of the Ishihara test for 'unlettered persons', more efficient screening is achieved by using adult tests and adapting the examination method to the ability of the child. Specially designed pseudoisochromatic tests for children and adult tests which are easily understood by children are listed in Table 8.2.

Failure to respond correctly to the introductory plate means that the test cannot proceed. Occasionally, children with psychological problems attempt to simulate colour deficiency. In these rare instances, the examiner is alerted by failure of the introductory plate or inconsistencies in the child's responses. For example, plates containing large colour differences may be failed while those containing small colour differences are interpreted correctly. Mixed protan, deutan, and tritan results may be reported. Referral to a medical practitioner is necessary in these cases.

Testing up to 7 years of age

The visual pathway is not fully developed until about 3 months of age. From 3 months onwards colour vision can be examined objectively by measuring chromatic ERGs or by measurement of the pupil response to coloured targets. Subjective examination can be attempted from about 3 years of age.

Table 8.2 Pseudoisochromatic tests for children

Test	Format	Replica figure provided to aid identification
Ishihara plates	Pathways	No
Ishihara plates for 'unlettered persons'	Symbols and pathways	Yes (symbols)
Guy's colour vision test for children	Upper-case letters	Yes
Velhagen Pflüger-trident plates	Illiterate 'E'	Yes
Ohkuma plates	Landolt ring	Yes
Matsubara plates	Pictograms	No
American Optical Co. (HRR) plates	Symbols	No

Games

Colour games consisting of counters painted with isochromatic colours which have to be matched, as in 'lotto' or 'domino', have been used to examine 3-year-olds (Verriest 1981). In the Fletcher–Hamblin simplified colour vision test, isochromatic colours are presented in a pattern. The test consists of three mosaic tiles. The child has to point to colours which look different from the majority of the mosaic in the first two tiles. In the third tile the child selects the colour which looks most like a reference colour. Ceramic is used so that the tiles can be handled and wiped clean. At the end of the test, the examiner must decide whether incorrect results are due to abnormal colour vision or to poor understanding of the visual task. This usually necessitates playing the game again in a slightly different way and several repetitions may be necessary to confirm colour deficiency. The colours used in games must have fairly large colour differences otherwise children with normal colour vision would find them too difficult: in consequence only children with severe colour deficiency are likely to be identified.

Preferential looking

Preferential looking, with individual pseudoisochromatic designs as the test target, can be used for examining babies and toddlers. Pease and Allen (1988) have developed a test on this principle. The pseudoisochromatic design is based on the colours used in the Farnsworth F2 plate and consists of a single square positioned to one side of a rectangular background. There are four plates. Two plates are for demonstrating the visual task and are used to confirm that the child is responding correctly. One plate is for detecting red–green defects and one plate for tritan defects. Each plate is shown several times, with the square positioned either on the right or on the left side of the design. A number of repetitions are needed.

Optokinetic nystagmus has also been used to assess infant colour vision. Isochromatic colour stripes of equal luminance are painted onto a drum. Tracking eye movements when the drum is rotated show that the infant can distinguish the colours and absence of eye movement suggests colour deficiency of a particular type.

Pseudoisochromatic plates

Colour games and preferential looking techniques take a long time to complete and are not suitable for examining large numbers of children. Pseudoisochromatic plates are easily understood and have a much greater potential for screening large groups. However, most pseudoisochromatic tests for adults contain numerals which young children have difficulty recognizing and naming. Specially designed children's tests either contain simple shapes, or pictures, which are easier for a child to name, or provide printed or cutout replicas which the child selects to match the one seen if it cannot be identified verbally. Examiners can make their own replicas for individual plates of standard tests or can encourage the child to draw over the figure using a clean paint brush or cotton bud.

However, the longer viewing time necessary to draw over the design reduces screening accuracy and children with slight colour deficiency may not be detected. Children with normal colour vision find it difficult to draw over transformation designs because the longer viewing time enables them to see elements of both the normal and confusion figures and an ambiguous result is often obtained. A small number of plates are usually shown so that the child's attention span is not exceeded.

The Velhagen Pflügertrident test contains a letter 'E' format and the child has to turn a hand-held replica to match the orientation of the figure seen. Children over 5 years of age enjoy this task but younger children find it difficult to manipulate the 'E' and point instead. All the plates are of the vanishing type and three responses are possible; the correct result, incorrect orientation of the replica, and failure to see the 'E'. Children with normal colour vision under 7 years of age frequently mismatch the 'E' and colour deficient children frequently position the replica incorrectly rather than say that they cannot see the 'E'. These responses make it very difficult to identify colour deficient children with this type of test. In addition, plate 9 is a low threshold plate and cannot be distinguished by either normal or colour deficient children.

The Ohkuma pseudoisochromatic test employs a similar examination method but contains a Landolt C. However, the colour design is not correctly realized and poor results have been obtained with adult subjects (see Chapter 7).

The Matsubara test contains pictograms, which the child has to name, and the Guy's colour vision test contains upper-case letters. The Guy's test includes replicas which the child selects to match the letter seen. Both the Matsubara test and the Guy's test have poor colour design and a large number children with normal colour vision fail both tests. Screening is therefore unreliable. The Guy's colour vision test has 8 plates; 2 plates are for instruction, 4 plates are for red–green screening, and 2 plates for protan/deutan classification. The plates have a glossy surface which can be wiped clean if handled carelessly. The Matsubara test consists of 10 plates. The first plate is for demonstration, 5 plates are for red–green screening, and 4 plates for protan/deutan classification. All the plates are of the 'vanishing' type and a different pictogram is used on each. Some of the pictures, such as the tortoise, are quite difficult to recognize. The instructions provided with the test are confusing and the information given about colour deficiency is incorrect.

An Ishihara test for 'unlettered persons' is an excellent screening test for young children and takes only 3 or 4 min to complete. The test became available in the UK in 1990 although it was apparently available in Japan in 1970 (Verriest 1981). The test is intended for use with 4–6 year-olds. There are 8 plates. Four plates contain simple shapes, either a circle or a square, and 4 plates contain pathways. The pathway designs are less complex than those of the standard Ishihara test. Two of the four symbol designs are for introduction and 2 plates are confusion designs containing either a circle or a square. These two designs are based on the most efficient colour combinations used in the standard Ishihara plates and have the same accuracy.

One of the pathway designs is for introduction, 2 plates are confusion designs, and one plate is for protan/deutan classification. The classification plate is the same as plate 26 of the standard Ishihara test. Unfortunately children with normal colour vision can see both pathways on plate 6 and, although the design works correctly in distinguishing colour deficiency, it is better not to include it in the examination. Similarly, there is no need to show the classification plate (plate 8) containing two pathways unless colour deficiency is found. The screening test therefore consists of 6 plates only, 3 are for demonstration and 3 for the examination. The test can be used with children under 4 years of age but the plates may have to be shown twice to confirm the first result if the child is hesitant or makes an error. Children over 4 years of age usually complete the test first time without requiring any additional aids to identification. However circles are often described as 'rings' or 'oh's'.

Children under 7 years of age can be examined with standard pseudoisochromatic tests if a small number of carefully chosen plates are given and replicas made available. Tracing can be encouraged as an alternative to using replicas. A recommended selection of six of the most efficient designs from the Ishihara plates is shown in Table 8.3. These plates require the identification of only four numbers; 1, 2, 3, and 5. Two plates, the introduction and classification plates, contain pairs of numerals but a card can be placed over half of the plate so that only one figure need be identified at a time. Two plates (plates 6 and 7 of the standard test) are 'transformation' designs and two are 'vanishing' (plates 10 and 14). Children respond best to numbers up to their chronological age and plates containing larger numbers should be avoided. The replica of a 6 is often inverted by young children and plates containing

Table 8.3 Selection of six plates from the Ishihara pseudoisochromatic test for screening children for red–green deficiency

Plate	Numerals contained		Design	Responses Normal		red–green deficient	
1	1	2	Introduction	1	2	1	2
6	2	5	Transformation	5		2	
7	5	3	Transformation	3		5	
10	2		Vanishing	2		—	
14	5		Vanishing	5		—	
24	3	5	Classification	3	5	Protans read 5 Deutans read 3	

6's and 9's should be avoided. About 50 per cent of 3–4 year-olds choose to trace over the numbers and the plates have to be shown three times in about 25 per cent of examinations. A small number of children in the 5–7 year age group choose to select matching replicas but usually only one showing of the plates is necessary. The test takes about 5 min to complete or 15 min if three repetitions are needed. The test takes longer and requires more expertise from the examiner than the Ishihara test for 'unlettered persons' especially in the youngest age groups.

The standard Ishihara plates and the Dvorine plates contain designs with pathways which are intended for use with nonverbal subjects. The designs are more complex than those included in the Ishihara test for 'unlettered persons' but a small number of plates containing transformation and vanishing designs can be selected to examine children if required. Tracing over the pathways takes a little longer and screening efficiency is reduced in consequence.

The Ishihara test is only useful for screening and gives no indication of the severity of colour deficiency. If an estimate of the severity of colour deficiency is needed, either the American Optical Company (HRR) test, the Farnsworth D15 test, or the City University test can be given as a second test. Failure of one of these tests identifies severe colour deficiency (see Chapter 7). The HRR plates should be shown in reverse order, so that the test begins at an easy rather than a difficult level. Clear diagnostic responses, in which one of the paired symbols is detected and not the other, is a clear demonstration of severe colour deficiency. The Farnsworth D15 test can be used successfully if a 'step-by-step' procedure is adopted. The idea of making a natural colour order is generally too difficult for children in this age group and the examiner merely asks the child to select the colour which looks most like the last colour in the box after each selection has been made. The advantage of this test method is that the examiner can observe isochromatic colours being considered even if these are not eventually placed together in the arrangement. The

City University test was developed specially for the examination of children and is the preferred test for grading the severity of colour deficiency in young children. The test contains the same colours as the D15 test but is in book form so that manual dexterity is not needed. The child points to the colour which looks most like the test colour in each design. Plates 7–10 of the second edition contain small desaturated colours and are too difficult for children and only plates 1–6 are used. The idea of choosing the most similar colour is not easy for young children and the instructions may have to be repeated for each plate. Protan/deutan classification is not always reliable with this test and mixed protan/deutan responses frequently occur as an artefact of the test design.

Testing children from 7 to 11 years of age

Between the ages of seven and eleven years it is possible to administer standard pseudoisochromatic tests in the manner of an adult examination and the examiner can be confident that accurate colour vision screening has been achieved. Most pseudoisochromatic tests aim to distinguish protan and deutan defects and some tests also screen for moderate tritan defects. The Ishihara test, the HRR test, the D15 panel, and the City University test can all be given in the normal way.

Few children under the age of 10 years have the necessary cognitive skills to complete the F–M 100 hue test. This test is more appropriate for examining young people over 15 years of age and is most useful when career options are being considered.

Periodic health screening in schools affords the opportunity to repeat the colour vision assessment on each occasion. A repeat examination at about 11 years of age confirms the first screening results and is an appropriate time to mention career restrictions. Detailed careers advice can be delayed until the final medical examination at 15 years of age.

References

Pease, P.L. and Allen, J. (1988). A new test for screening color vision. *American Journal of Optometry and Physiological Optics*. **65**, 9, 729–38.

Verriest, G. (1981). Colour vision tests in children. *Atti. Fondazione Giorgio Ronchi*, **36**, 83–119.

Voke, J. (1976). The industrial consequences of deficiencies of colour vision. Ph.D. thesis. The City University, London.

9. Acquired colour vision defects

Defective colour vision can be acquired as a result of ocular pathology, intracranial injury, or by the excessive use of therapeutic drugs. Unlike congenital colour deficiency, which arises from abnormal photochemistry, acquired defects can be caused by abnormalities anywhere in the visual pathway from the retinal receptors to the visual cortex. Changes in colour vision give evidence of these abnormalities. Identification of a particular type of colour deficiency with a specific pathological lesion provides information on how colour is processed in the visual pathway. Colour disturbance is an early symptom of some conditions and occurs at a recognized stage in others. It follows, therefore, that detection of acquired colour deficiency is an important diagnostic aid and may suggest when therapeutic measures should begin or when they need to be discontinued. Changes in colour vision are frequently used to monitor ocular pathology and to assess treatments.

Congenital and acquired colour deficiency differ in several ways (Table 9.1). Congenital colour deficiency is stable throughout life and both eyes are equally affected. Acquired colour vision defects change in severity with time and monocular differences in severity are often found. Colour deficiency may vary sectorially in an affected eye. Acquired defects differ qualitatively. They are less easy to classify, and nonspecific or combined defects occur which have characteristics associated with more than one type of congenital colour deficiency. Other changes in visual function are found, including reduction in visual acuity, abnormal contrast sensitivity, and visual field defects. If acquired colour deficiency is suspected, the colour vision examination is always carried out monocularly.

The first report of a person with acquired colour deficiency was made by Robert Boyle in 1688. This concerned a lady suffering from an unidentified illness who temporarily became blind, perhaps due to the 'fierce vesication' of the treatment she received. Her vision gradually returned but with a complete absence of colour vision, a situation which is now known to occur following intracranial injury. In a charming passage,

Table 9.1 Characteristics of congenital and acquired colour deficiency

Congenital colour vision defects	Acquired colour vision defects
1. Present at birth	Onset after birth
2. The type and severity of the defect is the same throughout life	The type and severity of the deficiency fluctuates
3. The type of defect can be classified precisely	The type of defect may not be easy to classify. Combined or non-specific defects frequently occur
4. Both eyes are equally affected	Monocular differences in the type and severity of the defect frequently occur
5. Visual acuity is unaffected (except in monochromatism) and visual fields are normal	Visual acuity is often reduced and visual field defects frequently occur
6. Predominantly either protan or deutan	Predominantly tritan
7. Higher incidence in males	Equal incidence in males and females

Boyle described how she wanted to pick violets but could not distinguish them by colour from the surrounding leaves. This report is particularly interesting because it suggests a tritan defect which is now known to occur in some types of intracranial injury. In the nineteenth century acquired colour deficiency received the same attention as that given to congenital defects and several cases of 'amnestic colour blindness' resulting from cerebro-vascular lesions were reported. Complete loss of colour perception caused by a concussion injury, sustained in a riding accident, was described by Wilson in 1855.

Helmholtz published his description of a prototype ophthalmoscope in 1851 and by the end of the next decade ophthalmoscopic examination of the ocular fundus had become routine. Ophthalmoscopy revolutionized the differential diagnosis of ocular pathology and it was possible to show that particular diseases caused specific colour vision abnormalities. The association of acquired red–green blindness with tobacco amblyopia, and acquired tritanopia with retinal detachment was described by Konig in 1897 and the first classification of acquired colour deficiency was made by Köllner in 1912. Köllner accepted the Hering theory of colour vision and, because of the terms used, Köllner's 'rule' is often misinterpreted as stating that 'Defects of blue–yellow vision are caused by retinal disease and defects of red–green vision are due to optic nerve disease.' In fact Köllner describes a progression of colour vision loss.

Blue–yellow blindness: Blue and yellow change their appearance first, green and red are preserved longer. Acquired 'blue–yellow blindness' especially develops in diseases of the retina and total colour blindness only results in combination with 'progressive red–green blindness'.

Progressive red–green blindness: Colour vision is totally disturbed. Blue–yellow vision is changed but deterioration is most striking for red and green. This type of colour blindness can especially be found in diseases of the conductive pathways reaching from the inner layers of the retina to the cortex.' (Marré 1973)

The results of modern examination techniques show that there are two principle exceptions to Köllner's 'rule'. Red–green defects are found in central retinal dystrophies and pathology involving the visual pathway above the chiasma produces acquired type 3 (tritan) colour deficiency or achromatopsia. The development of modern electrodiagnostic techniques and fluorescein angiography has led to more precise clinical diagnosis than was available to Köllner and it is possible to link acquired colour deficiency with types of neural or cellular pathology rather than to a broad classification of disease.

The same psychophysical methods as those for congenital colour deficiency can be used to analyse acquired defects but the consequences of reduced visual acuity and the quality of fixation have to be taken into account.

Measurements of relative luminance efficiency enable two types of acquired red–green deficiency to be distinguished (Table 9.2). In type 1 acquired red–green defects the wavelength of maximum relative luminous efficiency is displaced towards shorter wavelengths and in type 2 defects the wavelength of maximum sensitivity is unchanged. Type 1 defects are found in central retinal dystrophies and type 2 defects in some lesions of the optic nerve. Pathological conditions which are localized to the macular area destroy the central cone receptors. This causes loss of visual acuity and a Purkinje-like shift in the wavelength of maximum sensitivity, similar to that found in congenital protan defects (an acquired type 1 defect). This alteration is sometimes referred to as 'scotopization' of colour vision because the responses of the rod receptors become dominant giving enhanced sensitivity to short wavelengths and reduced sensitivity to long wavelengths. Scotopization is also found in extra foveal fixation. Optic nerve pathology involves loss of red-green opponent fibres, an overall loss of spectral sensitivity results but there is no significant change in the wavelength of maximum sensitivity. The resulting relative luminous efficiency resembles that found in congenital deutan deficiency (an acquired type 2 defect). Central visual field defects may be anticipated in type 1 and type 2 acquired colour deficiency.

Acquired type 3 (tritan) defects occur in many pathological conditions affecting both the central and peripheral retina and in lesions of the visual pathway above the chiasma. Type 3 defects closely resemble congenital tritan defects and acquired dichromats are able to match violet and yellow wavelengths as in congenital tritanopia. However, the relative luminous efficiency differs from that of congenital tritanopes. There is a much greater loss of sensitivity at short wavelengths together with loss of sensitivity at the long-wave limit of the spectrum. This changes the perceived luminance contrast between pairs of isochromatic colours and

Table 9.2 Classification of acquired colour deficiency

Classification	Characteristics	Association
Type 1 red–green	Similar to protan deficiency—displaced relative luminous efficiency to short wavelengths	Progressive cone dystrophies Retinal pigment epithelium dystrophies
Type 2 red–green	Similar to deutan deficiency but with greater reduction in sensitivity to short wavelengths	Optic neuritis
Type 3 tritan (blue)	(a) Similar to tritan deficiency but with displaced relative luminous efficiency to short wavelengths	Central serous chorioretinopathy Age-related macular degeneration
	(b) Similar to tritan deficiency	Rod and rod–cone dystrophies Retinal vascular disorders Peripheral retinal lesions Glaucoma Autosomal dominant optic atrophy

pseudoisochromatic designs may not be equally effective for detecting congenital and acquired tritan deficiency. Peripheral field loss is associated with acquired type 3 defects.

Type 3 defects are often indiscriminantly referred to as 'blue–yellow' in the literature in accordance with the nomenclature proposed by Köllner. A neutral zone is found in the yellow region of the spectrum in both congenital tritan and type 3 defects, and may be considerably enlarged in the latter. However, the term 'blue–yellow' incorrectly implies absence of a 'yellow' receptor and suggests that blues and yellows are confused. This is not so. Typical colour confusions are between blues and greens, or between yellows and violets, never between blue and yellow. In addition, type 3 defects are sometimes incorrectly described as showing a 'blue–yellow axis' on the F–M 100 hue test when the characteristic reduction in hue discrimination occurs in the red–violet and blue–green parts of the hue circle in both congenital tritan defects and in acquired type 3 defects. 'Blue–yellow' is only appropriate for the stage in type 3 defects when all three colour vision mechanisms, (red, green, and blue), are significantly affected and the term should be limited to this context.

The measurement of threshold spectral sensitivity has considerably advanced the understanding of acquired colour deficiency. Measurement of colour vision mechanisms, using the Wald–Marré technique of selective adaptation, shows that two different stages of colour deficiency occur in retinal disease. These are the same irrespective of the causative pathology or the location of the lesion (Marré 1973). Initially there is an isolated type 3 defect but as the disease progresses all three colour vision mechanisms become abnormal (a 'blue–yellow defect'). Significant involvement of the red and green mechanisms is never found without complete loss of the blue mechanism. The red and green mechanisms are equally affected and become reduced at the same stage of the disease. Gradual loss of all three colour vision mechanisms explains the reduction in long-wave sensitivity found in type 3 defects.

Three types of acquired colour deficiency are found in diseases of the optic nerve using the Wald–Marré technique. These are: (a) an isolated disturbance of the blue mechanism; (b) combined disturbance of the blue and green mechanisms; and (c) combined disturbance of all three colour vision mechanisms. Isolated type 3 defects occur briefly in the initial stages of optic nerve pathology. Disturbance of blue and green mechanisms is a brief transitional stage leading to combined colour deficiency which is the most typical acquired defect in optic nerve disease. Reduction in all three colour vision mechanisms occurs at a much earlier stage than in retinal disease and is found when visual acuity is relatively good.

Measurements of threshold spectral sensitivity therefore show that all acquired defects progress in stages. There is an abnormal trichromatic stage, then a form of dichromatic vision, and finally complete loss of colour vision or achromatopsia. Type 3 defects always occur first in the initial stages of both retinal and optic nerve pathology but the duration may be very brief before the characteristic type 1 and type 2 defects develop. Measurement of visual acuity and visual field loss help to identify these stages.

Lesions in the optic pathway and visual cortex cause a variety of sensory dysfunctions including chromatopsia (coloured vision) and achromatopsia (complete colour blindness). Chromatopsia is a transient phenomenon in which the environment appears suffused in colour. Specific terms are used to describe the colour seen. 'Cyanopsia' describes a blue sensation; 'chloropsia', green; 'xanthopsia', yellow; and 'erythropsia', red. Achromatopsia of cortical origin is associated with occipital lesions which produce upper bilateral visual field loss and other sensory deficits such as impaired topographical memory, colour anomia, and prosopagnosia. Colour anomia is the inability to remember colour names and prosopagnosia the failure to recognize familiar faces.

Clinical examination procedure

A test battery is essential for the evaluation of acquired colour deficiency and each eye is always examined separately. The tests recommended are the same as for congenital colour deficiency except that the F–M 100 hue test is obligatory and a lantern test is omitted.

Pseudoisochromatic tests for tritan defects must be included in the test battery and the addition of the Besançon anomalometer, or an adaptation test, such as the TNO test, for examining threshold blue sensitivity, is desirable. A graded series of D15 tests is very useful.

The analysis of acquired colour deficiency is not always straightforward. If eccentric fixation is present, the test results will correspond with extrafoveal vision and must be compared with extrafoveal colour vision characteristics in normal trichromats. Reduced visual acuity and visual field defects restrict the use of some clinical tests and detection of type 3 defects in older patients is complicated by age-related changes in short-wavelength sensitivity. Cataract and vitreous opacities are frequent secondary features in retinal pathology and change threshold blue sensitivity to a greater extent than in normal trichromats of the same age. This may invalidate comparisons between patients and aged-matched normal observers. Detection of an acquired defect is more difficult if the person has congenital colour deficiency. The characteristics of the two types of deficiency are combined and it is impossible to monitor changes in hue discrimination to assess the underlying pathology. A congenital red–green dichromat who acquires a type 3 defect becomes virtually monochromatic.

A strict protocol is used in the examination of congenital colour deficiency but a more flexible approach may have to be adopted for acquired defects. Patients who have sensory deficits may not be able to respond to standard clinical tests and simple items such as coloured cards, felt-tip pens, and children's colouring books are useful. Colour naming is usually avoided in the examination of congenital colour deficiency but can provide fascinating insight into the nature of acquired colour deficiency. In some cases the patient can give a detailed description of colour vision changes and can compare colours seen with an affected and nonaffected eye.

Pseudoisochromatic plates

Poor visual acuity and paracentral field defects affect the persons ability to complete pseudoisochromatic plates. Some pseudoisochromatic tests are more robust than others to changes in visual acuity (see Table 6.3, pp. 58). For example, the acuity limit for the Ishihara plates is about 6/18 but about 6/60 for the HRR plates. The critical value for visual acuity is determined by the size elements of the design and tests cannot be used effectively if acuity is below this value. Observation of pseudoisochromatic designs is a dynamic visual task and people scan the plate in order to search for the contained figure. This means that the figure is partly constructed from sequential observations. Paracentral field loss may interfere with this construction and adversely affect performance even when acuity is better than the critical value. It is therefore essential that both visual acuity and central fields are checked before the examination begins. Amsler charts are sufficient to confirm the integrity of the central 10° of visual field. Amsler charts consist of a small square grid of lines; the patient views a central fixation mark and reports any distortion or discontinuity of the lines outside the fixation area.

Typical results are obtained for the Ishihara plates in type 1 and type

2 defects. In the initial stages the patient responds slowly and an increased number of interpretive misreadings are recorded. As the deficiency becomes more severe, vanishing errors as well as misreadings occur. The patient cannot distinguish the numeral on vanishing designs and neither the correct nor the confusion numerals can be seen on transformation designs (plates 2–9). This response is an important difference between congenital and acquired red–green colour deficiency. Similarly, neither of the protan/deutan classification numerals may be distinguished (plates 22–25). This response is sometimes obtained in congenital colour deficiency, if a high density of macular pigment is present, but visual acuity is normal, whereas in acquired type 1 or type 2 defects acuity is always reduced. In the final stages, when acuity reaches the limit of the test design, only the high contrast introductory plate can be seen. Patients with severe type 3 defects respond in a similar way when the red/green mechanism is significantly involved. Acuity is usually within normal limits for these patients but visual field abnormalities are always present.

The HRR plates can be used when visual acuity is as low as 6/60. The severity of the red–green defect can be estimated using the grading plates but, as with the Ishihara plates, a differential diagnosis of either protan (type 1) or deutan (type 2) is not usually obtained. Moderate and severe type 3 defects are identified by the HRR tritan designs. However, the small colour differences used in the desaturated red/green screening plates are difficult for people with acquired type 3 defects to interpret and a concomitant red–green defect is recorded as an artefact of the test design before the red and green mechanisms are significantly reduced. Similar small colour differences are utilized in the SPP 2 plates and mixed tritan and red/green diagnostic responses are made at an early stage. These results limit the use of these particular tests for identifying, and grading, the severity of acquired tritan defects. Involvement of the red/green mechanism in type 3 defects is more clearly shown by the results of the Ishihara plates and the F–M 100 hue test. A clear tritan axis of confusion is obtained with the F–M 100 hue test in type 3 defects before involvement of the red–green mechanism. When the red–green mechanism is compromised the result shows poor overall hue discrimination. For example, in diabetic retinopathy a direct relationship is found between the F–M 100 hue error score, the presence of an axis of confusion, and the size of the blue mechanism measured with the TNO test (Fig. 9.1). When the defect is severe, the error score is high, there is no discernable axis of confusion, and the blue mechanism is functionally absent. Acquired tritanopia is confirmed by the patient's ability to make tritanopic colour matches, between violet and yellow wavelengths.

The F2 plate is ineffective for detecting acquired type 3 defects and only severe defects are identified by plate 5 (the plate with the lowest colour difference threshold) of the Lanthony tritan album. About 35 per cent of patients with known acquired type 3 defects are identified with the Lanthony tritan album. The new City University tritan plates are more successful in detecting and grading acquired tritan defects. The test contains designs with small colour differences which are intended for screening, and designs with larger colour differences to distinguish severe defects.

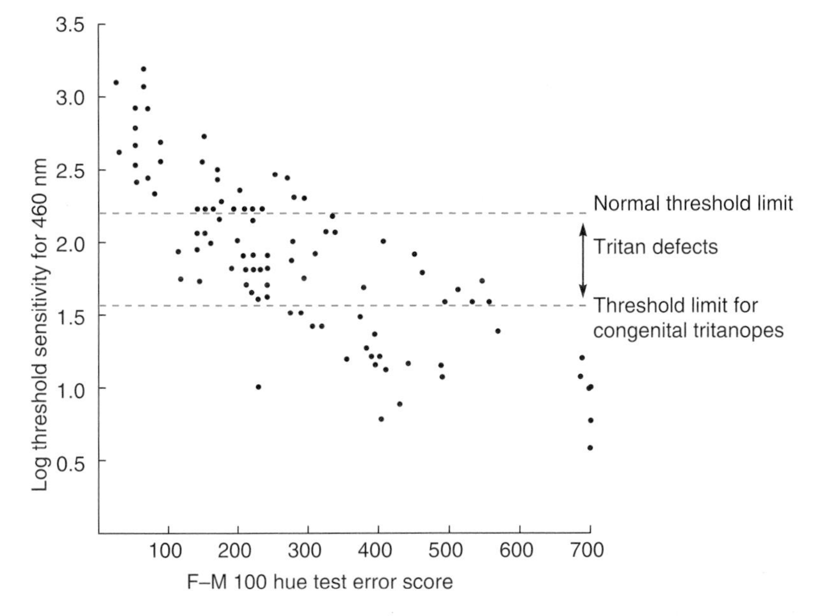

Fig. 9.1 Threshold blue perception at 460 nm measured with the TNO test and the F–M 100 hue error score for 85 patients with diabetic retinopathy. In acquired type 3 (tritan) defects, short-wave sensitivity at 460 nm is reduced more than in congenital tritanopia. This is due to involvement of the red–green mechanism which is only partially adapted by the measurement technique. Threshold blue sensitivity is inversely related to the F–M 100 hue error score even when no tritan axis of confusion is found.

Hue discrimination or arrangement tests

The hue discrimination tests designed by Farnsworth are widely used to study acquired colour deficiency. The F–M 100 hue test is particularly useful for showing the extent of the discrimination loss, for comparing the severity of deficiency in each eye, and for monitoring changes with time (Fig. 9.2). In acquired colour deficiency the total error score is often large and the polar diagram is difficult to interpret. Several methods for analysing F–M 100 hue plots have been described. Averaging methods, which filter poor overall hue discrimination from the data in order to reveal the presence of an axis of confusion, are most useful for the study of acquired colour deficiency. The averaging method described by Dain and Birch (1987) retains the performance characteristics of the test and does not impose a cyclic variation on the test results. The error value for each cap of the test S_c is derived by comparing an individual error score with the mean error score of the ten preceding and ten following caps plus that of the cap itself. This value is then multiplied by the mean error score of the test.

$$S_c = \frac{\sum_{x=c-10}^{x=c+10} A_x}{21} \cdot \frac{\sum_{x=1}^{x=85} A_x}{85} \qquad (9.1)$$

where A_x is the actual score for cap x.

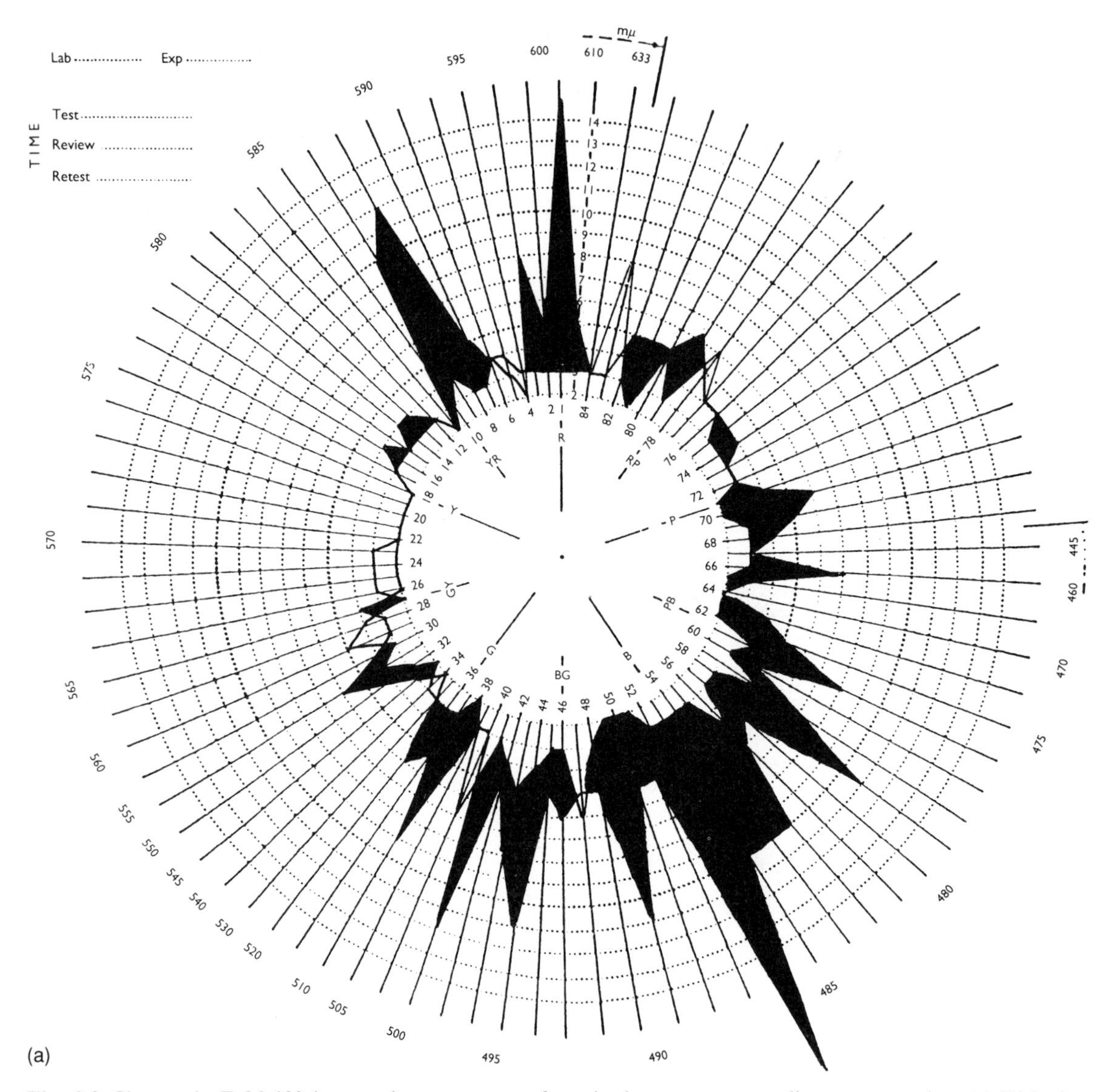

Fig. 9.2 Changes in F–M 100 hue results as a means of monitoring treatment or disease progression. (a) Diabetic retinopathy. A tritan axis of confusion found following pan retinal photocoagulation with the argon laser. The error score increased from 100 (before treatment) to 268 (dark pattern) afterwards. (b) Retrobulbar neuritis associated with multiple sclerosis. Spontaneous recovery of a type 2 red–green defect accompanying improvement of visual acuity from 6/12 to 6/9 after 1 month. The error score is reduced from 316 to 192 (dark pattern). (c) Vitamin A deficiency associated with billiary cirrhosis (Bronté-Stewart and Foulds 1971). Recovery of nonspecific hue discrimination loss following oral dose of vitamin A and improvement of visual acuity from 6/9 to 6/6. The error score was 244 before treatment and 96 (dark pattern) afterwards.

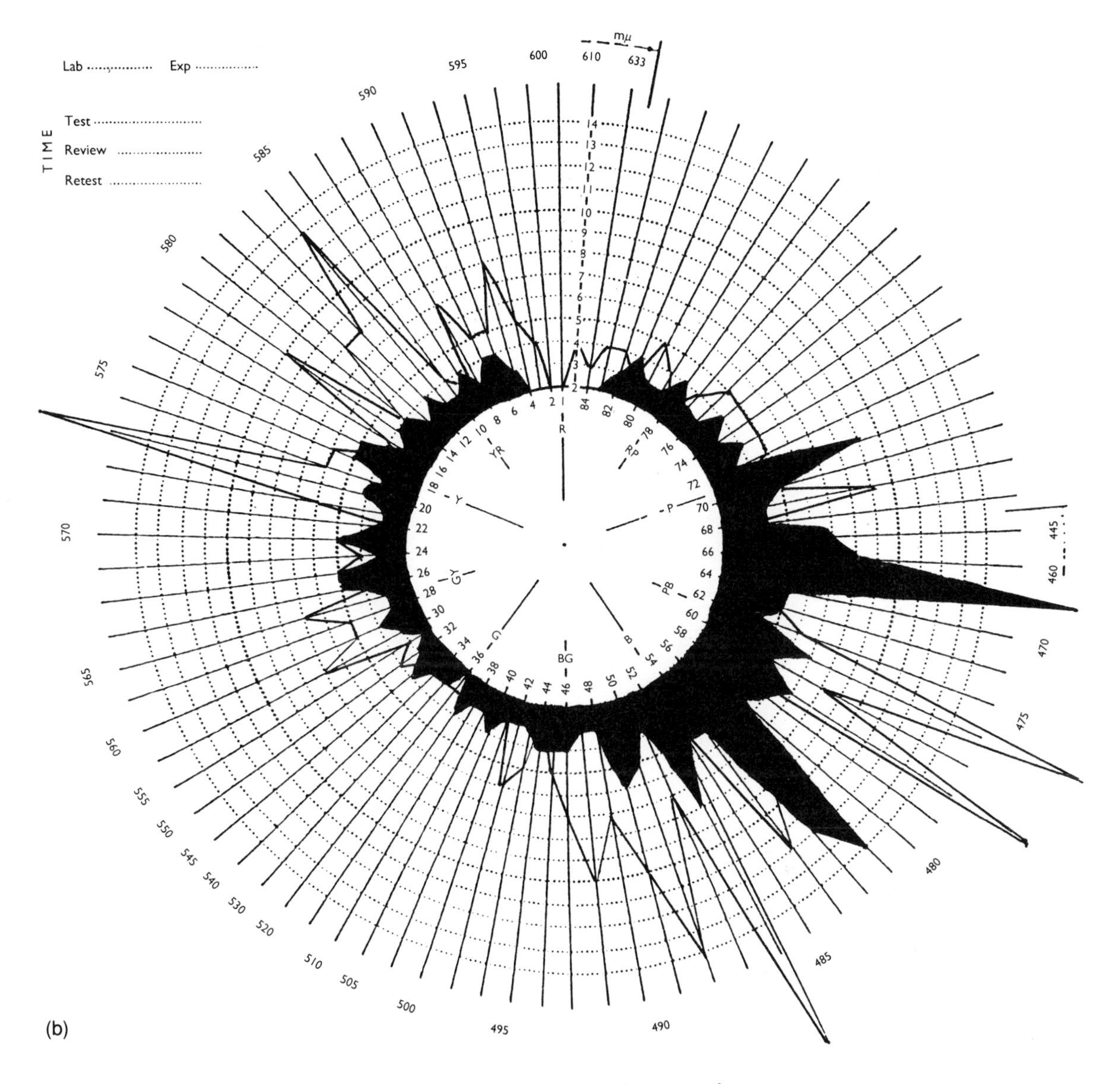

(b)

This method of calculation considers the score for each colour sample in relation to the particular quadrant of the test, for which it is the central colour, and compares this with the error score for the test as a whole. The analysis converts the results into a sinusoidal–like curve if an axis of confusion is present (Fig. 9.3). The amplitude of the curve shows the prominence of the axis of confusion and the height of the troughs above the x-axis show the amount of overall poor hue discrimination.

The F–M 100 test is not a reliable screening test for acquired colour deficiency. Attempts to use it for screening have relied on comparisons between the error scores obtained by patients and age-matched normal observers. Verriest's data is often used in this context (Table 6.4, p. 64) although a different level of illumination or test method may be employed. Although differences in the mean error scores are found between

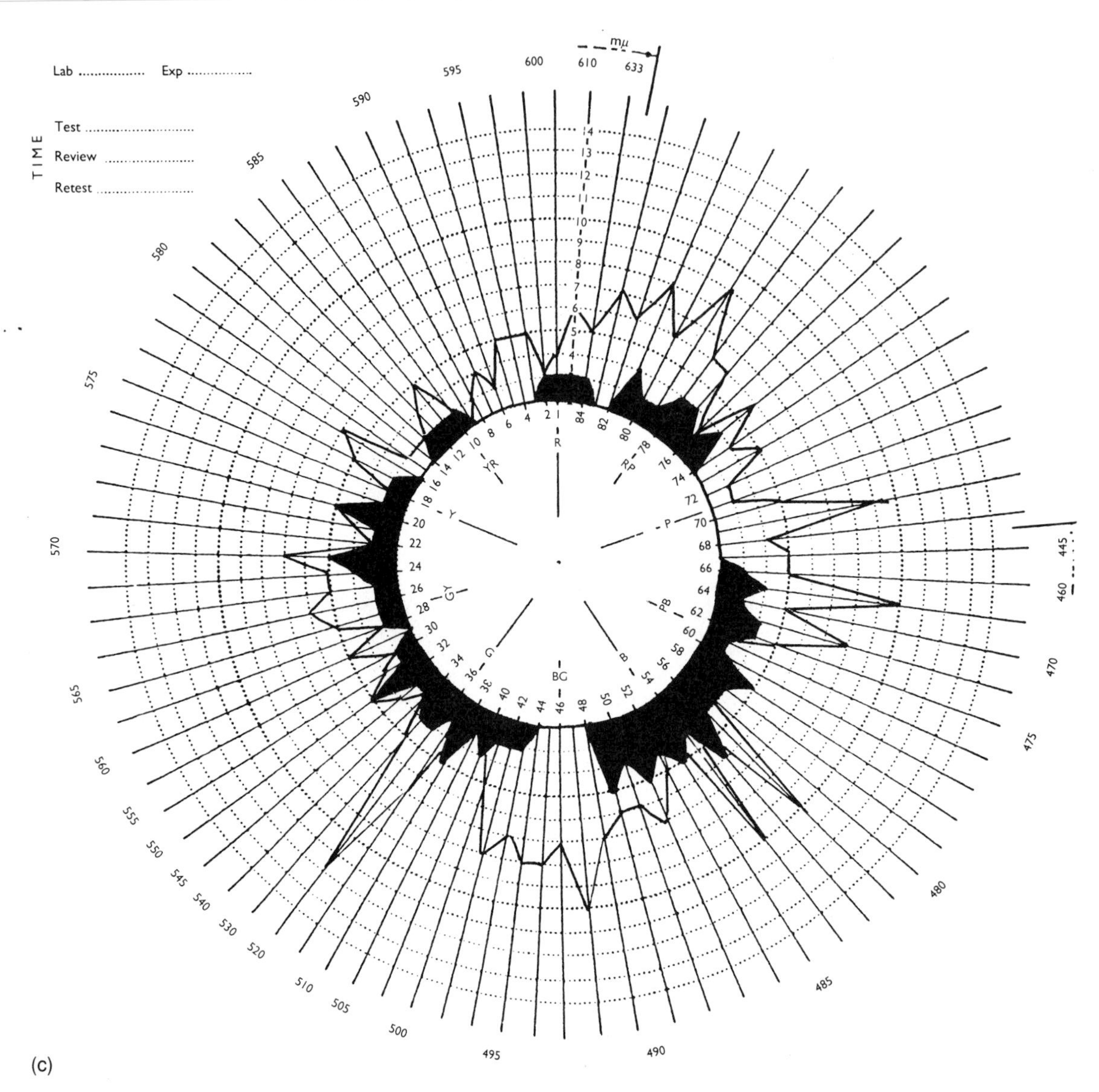

(c)

these groups, careful inspection of the data shows a considerable overlap which prevents this from being useful clinically.

The D15 test gives a rapid diagnosis of the type of acquired colour deficiency and a series of D15 tests, with different saturation levels, can be used to analyse the severity of the defect. In severe type 3 defects the D15 test may still give a clear tritan result at the stage when the F–M 100 test shows poor overall hue discrimination (Fig. 9.3). A version of the test having larger size colour caps has been shown to be effective for patients with poor visual acuity.

Scoring techniques have been developed for the D15 test but the error score is difficult to relate to the performance characteristics of the test. The colour difference step is not uniform across the hue circle and

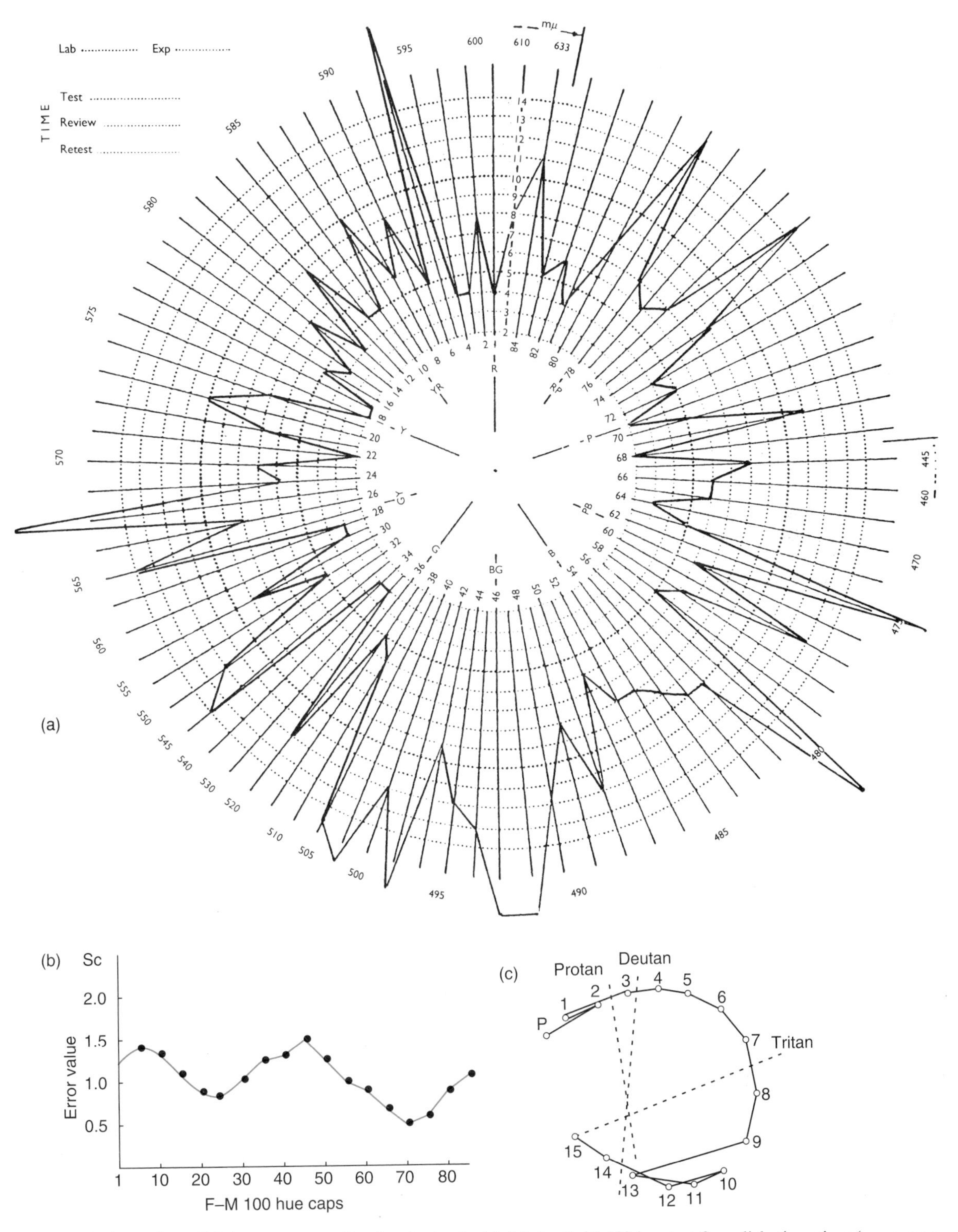

Fig 9.3 Averaging of high error scores. Results obtained with (a) the F–M 100 hue test for a diabetic patient (error score 588) reinterpreted using the averaging method (b) of Dain and Birch (1987) to show a type 3 defect superimposed upon poor overall hue discrimination. The D15 test (c) also showed a type 3 (tritan) defect for this patient.

isochromatic confusions are not of equal numerical value. A smaller number of possible isochromatic confusions are available to tritanopes than to people with red–green defects. Scoring systems based on the number of errors, or a calculation of the sum of the colour differences associated with those errors, often fail to provide more information than can be obtained from visual inspection of the record chart. A graded series of D15 tests is more useful for monitoring acquired defects than scoring a single test. The new color test (Lanthony) is similar to a series of D15 tests and can be used to demonstrate enlarged neutral zones. The City University test is not a suitable alternative to the D15 test for the examination of acquired colour deficiency.

Anomaloscopes

The Nagel anomaloscope can be used to evaluate acquired colour deficiency. The matching ranges obtained in type 1 and type 2 defects resemble those of protanomalous and deuteranomalous trichromats respectively, except that the matches are made at different luminance values and the matching range usually includes the normal matching range. The midpoint of the matching range determines whether the deficiency is type 1 or type 2. Extended matching ranges occur as a consequence of large neutral zones in the spectrum which are characteristic of acquired defects. Selective loss of the long- or medium-wave sensitive photopigments is never found in acquired colour deficiency but severe acquired red–green defects appear to reach a dichromatic stage in which all red/green ratios are accepted as a match. Type 3 defects caused by central retinal lesions give rise to a displaced matching range with a higher proportion of red than in the normal match. This is known as 'pseudoprotanomaly'. Symmetrical enlargement of the normal matching range is found in type 3 defects arising from peripheral retinal lesions.

Results obtained with the Besançon anomalometer, using the two-equation technique, illustrate the characteristics of the three types of acquired deficiency. People with type 1 colour deficiency obtain a protanomalous Rayleigh match and a normal Moreland match. In type 2 defects, the Rayleigh match is normal and an enlarged matching range is found with the Moreland match. Both matches are enlarged and outside the normal range in type 3 defects.

Thresholds obtained with the TNO test show a greater loss of sensitivity at 460 nm in type 3 defects than in congenital tritanopia. This is because the intensity of the yellow adapting field does not completely eliminate a small contribution from the red–green mechanism. Involvement of the red–green mechanism in type 3 defects reduces this contribution and a greater loss of sensitivity is recorded (Fig. 9.1).

Colour perimetry

Characteristic visual field defects identify the underlying pathology producing acquired colour vision deficiency. In most cases, sufficient documentation can be obtained from standard perimetric techniques using achromatic targets but, in some instances, the use of coloured targets is informative. In kinetic perimetry two different isopters or limits of visibility can be plotted for coloured targets. The first is at the detection limit, when the target is just perceived, and the second is at the recognition limit, when the colour of the target is correctly identified. The difference between these two measurements is related to the photochromatic interval. The position of the normal isopter for any target varies with the test conditions and identical parameters must be employed if results are to be compared. Only detection limits need be recorded to determine whether the visual field is normal or abnormal. Red targets are used more frequently than other colours because luminance contrast between the target and the background is significantly altered due to differences in relative luminous efficiency. The size of the normal isopter for a small white target is the same as that for a larger red target and the visual fields are of equal size. However, in many pathological visual fields the isopter for the equivalent red target is reduced giving a much smaller field: this is described as 'disproportion'. Disproportion is due to reduction in relative luminous efficiency at long wavelengths and is symptomatic of ocular pathology leading to acquired colour disturbance.

Measurement of both detection and recognition isopters have been recorded and the results tend to support Kollner's 'rule'. Patients with retinal diseases have a much smaller recognition isopter for blue, while in optic nerve lesions the recognition isopters for both red and green are smaller than normal and the photochromatic interval is enlarged. Traditional methods of colour perimetry became almost obsolete after facilities to examine peripheral colour vision using increment threshold techniques were introduced in bowl perimeters. Computer-controlled equipment, such as the Humphrey visual field analyser, has taken increment threshold measurements a further step forward in sophistication but the test procedure is lengthy and tiring for the patient.

Some patients, especially those with intracranial injuries, are only able to respond to simple methods of visual field examination. Apparent changes in hue and saturation may be reported when a hand-held target is moved across the visual field. Changes in the appearance of a target placed in different quadrants of the visual field can demonstrate sectorial colour deficiency in hemianopic field defects. A large red target positioned first on one side of the fixation point and then on the other will appear desaturated or darker on the affected side.

The Umazume–Ohta test was developed to investigate changes in colour appearance in the central 18° of visual field. The test consists of six plates and is administered in a similar way to Amsler charts. Each plate contains spots of colour positioned along lines radiating from a central point. The patient maintains central fixation and describes the position of spots in the pattern which appear to be different from the central colour. Neutral colours for type 1, type 2, and type 3 acquired deficiency are selected for each plate.

Acquired defects in retinal pathology

(a) Central retinal lesions

Type 1 red–green acquired colour deficiency occurs in progressive cone dystrophies and in retinal pigment epithelium dystrophies. These rare inherited diseases are characterized by an abnormal cone ERG, reduced visual acuity, central visual field loss, photophobia, and acquired colour deficiency. Several different types of dystrophy have been described and subclassifications are frequently made according to retinal appearance as well as from the electroretinogram. The mode of inheritance varies but is usually either autosomal dominant or autosomal recessive. The age of onset and the rate of progression vary in different disease entities. Stargardt's disease is typical of the recessive form. In Stargardt's disease, colour vision and visual acuity are affected in adolescence and colour vision is reduced to achromatopsia, with an acuity of less than 6/60, in the fourth decade. The severity of colour deficiency parallels the reduction in visual acuity and progresses in a definite pattern depending on the stage of the disease and the age of the patient. Slight type 3 defects have been reported in the initial stages of both dominant and recessive cone dystrophies, but significant colour deficiency is not usually found by clinical tests until acuity is less than 6/12. The F–M 100 hue test frequently shows a nonspecific red–green defect in which poor hue discrimination includes both protan and deutan axes (Fig. 9.4). Results of a test battery in a family with the dominant form of the disease illustrate this progression (Marré *et al.* 1989). Type 3 colour deficiency is found in the second decade and a type 1 defect in the third. Achromatopsia with acuity less than 6/60 occurs in the fourth decade (Table 9.3).

Type 3 acquired defects are found in macular lesions which originate at the level of Bruch's membrane. Causative pathology includes age-related macular degeneration and central serous chorioretinopathy. The incidence of age-related macular disease increases progressively from the fifth decade and is a major cause of visual loss in the elderly. Different types of lesion can be identified by fluorescein angiography. Metamorphopsia (visual distortion) may be reported, and confirmed with the Amsler charts, before visual acuity is significantly reduced and some patients describe transient episodes of chromatopsia, persistence of coloured after images, and differences in monocular colour appearance. Colour loss is severe in atrophic and exudative lesions even when visual acuity is about 6/12. Changes in colour vision can precede involvement of the fellow eye in unilateral cases.

Central serous chorioretinopathy is a self-limiting disease which occurs predominantly in males in the fourth decade. The retinal pigment epithelium becomes detached from Bruch's membrane and fluid accumulates in the intervening space and in the inner retina. Only one eye is usually affected. The first symptoms are metamorphopsia and changes in colour vision. A severe type 3 colour deficiency, with pseudoprotanomaly, occurs in the active stage of the disease and decreases in severity as oedema subsides. Improved hue discrimination, shown by reduction in the F–M 100 hue error score and axis of confusion, confirms that recovery has taken place. Slight colour deficiency may remain after the condition has resolved and a unilateral tritan defect in males of this age group leads to suspicion of a previous central serous episode.

(b) Peripheral retinal lesions

Type 3 (tritan) defects are found in many ocular conditions which affect the peripheral retina. The severity of the defect parallels visual field loss

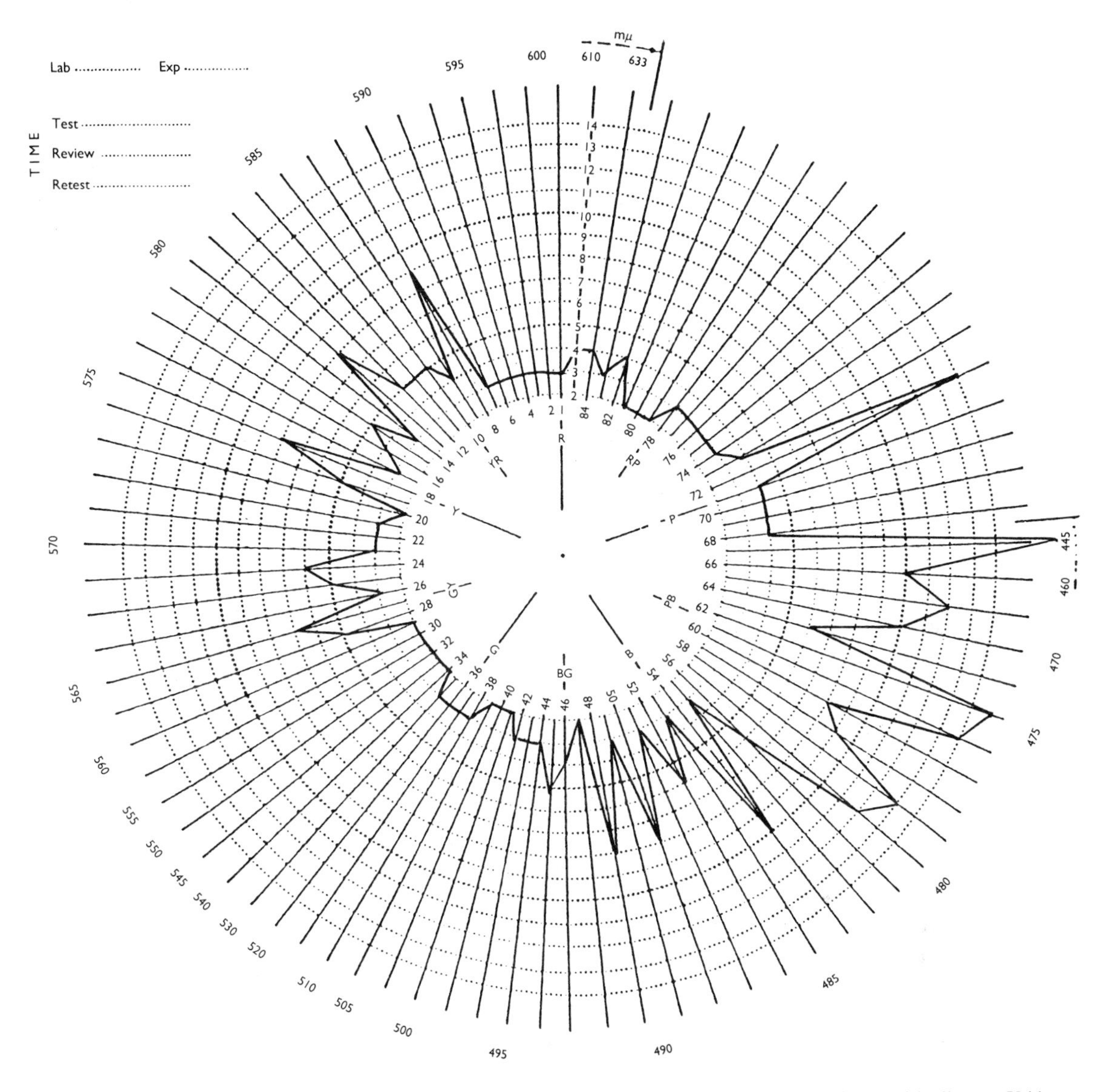

Fig. 9.4 The F–M 100 hue test in acquired colour deficiency. A type 1 red–green defect in Stargardt's disease, V/A = 6/18. Error score 248.

and severe colour deficiency may be present when visual acuity is within normal limits. Colour deficiency appears to arise from a 'whole retina response' but is most severe when both the peripheral retina and the macula are involved. The predominance of type 3 defects is a consequence of the spatial organization of the retina. A typical type 3 defect is found in some patients with retinal detachment, chorioretinitis, and hypertensive retinopathy. Initially there is an isolated disturbance of the blue mechanism and a tritan axis of confusion is obtained with the F–M 100 hue test. As the disease progresses all three colour vision mechanisms are affected.

Table 9.3 Colour vision results for four patients with autosomal dominant cone dystrophy in the same family

Patient	Age	V/A	Pseudoisochromatic plates SPP2	Pseudoisochromatic plates AO (HRR)	D15	Nagel anomaloscope	Wald mechanisms Red	Wald mechanisms Green	Wald mechanisms Blue
1	12	6/9	Pass	Pass	Pass	Pseudo protanomaly	34%*	25%*	62%*
2	13	6/18	Combined tritan and R–G defect	Slight tritan and slight R–G defect	Tritan defect	Protanomaly	69%	58%	1%
3	19	6/18	Combined tritan and R–G defect. Scotopic errors	Severe tritan and severe R–G defect	Tritan and R–G defect	Protanomaly	16%	20%	1%
4	36	6/60	Not possible	Not possible	Tritan and R–G defect	Full matching range with scotopic luminance values	3%	3%	0

* Doubtful reliability.
From Marré *et al.* (1989).

Retinitis pigmentosa

Colour vision is unaffected in the early stages of retinitis pigmentosa (RP). RP is a generic name for a group of inherited diseases which are characterized by night blindness and constricted visual fields. In the majority of cases there is no known family history and the inheritance is assumed to be autosomal recessive. Both autosomal dominant and X-linked variants of the disease are found. X-linked RP is the most aggressive form. Different genes which cause autosomal dominant RP have been localized on chromosomes 6 and 8, and abnormalities in the gene encoding rhodopsin have been discovered in other RP patients. The disease involves a breakdown of the phagocytic action of the retinal pigment epithelium and toxic cell debris accumulates in the retina. The ERG and the EOG are grossly abnormal. Both rods and cones are affected but damage to the rod system predominates. Central vision remains relatively intact in some forms of RP but in others the condition leads to complete blindness. A severe type 3 tritan defect, with pseudoprotanomaly, is found in types of RP which affect the central cones and may be a predictive factor for rapid loss of visual acuity. A broad tritan axis of confusion is obtained with the F–M 100 hue test (Fig. 9.5). The detection of acquired colour deficiency may initiate appropriate occupational advice and training before vision is severely compromised.

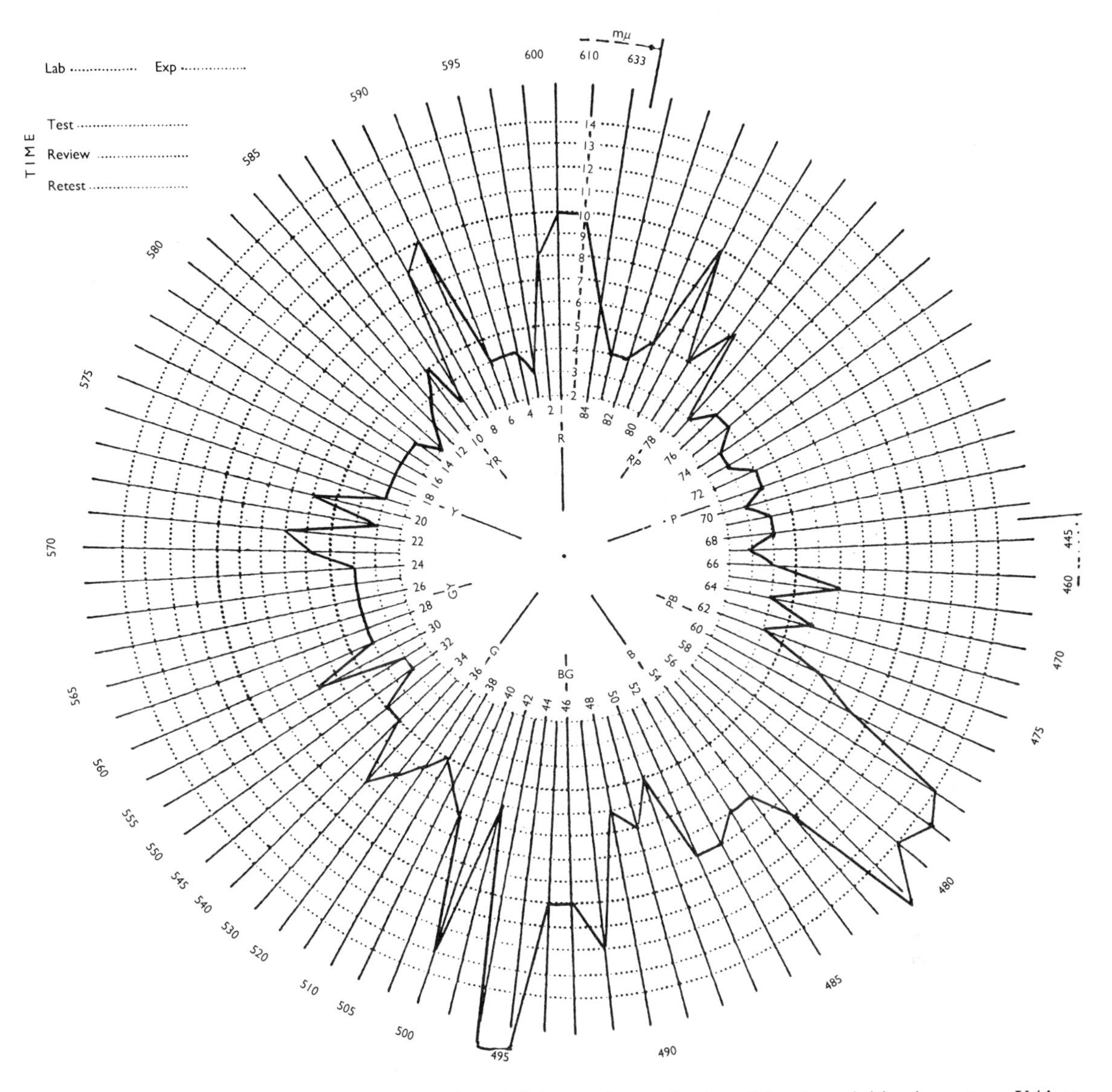

Fig 9.5 The F–M 100 hue test in acquired colour deficiency. A type 3 tritan defect in retinitis pigmentosa, V/A = 6/9. Error score 348.

Glaucoma

Typical type 3 (tritan) defects are found in glaucoma. Chronic simple glaucoma is estimated to affect about 2 per cent of the population over 50 years of age, and the disease is responsible for over 10 per cent of blind registrations in developed countries. There have been various attempts to use colour vision tests, especially the F–M 100 hue test and the Pickford–Nicolson anomaloscope, to predict the onset of field loss in ocular hypertensive patients and to show a relationship between colour disturbance and intraocular pressure. Performance on the F–M 100 hue test has been shown to deteriorate when intraocular pressure is raised artificially and a statistical analysis of F–M 100 hue error scores suggests that deterioration in hue discrimination precedes significant field loss. However, it is difficult to apply these data clinically due to normal variations in threshold blue perception, and differences in test performance, in the age group being examined. Visual fields and intraocular pressure measurements are more useful than colour vision testing for monitoring the disease. Field loss arises from diffuse changes in the neural retinal and the blue cone ERG is normal. These results show that acquired colour deficiency in glaucoma is caused by damage to the neural pathway rather than to the retinal receptors. The severity of the type 3 defect parallels field loss and acquired tritanopia is sometimes found.

Diabetic retinopathy

Type 3 (tritan) acquired colour deficiency is found in diabetic retinopathy. Diabetes affects approximately 2 per cent of the population worldwide and the number of diabetic patients is increasing by about 6 per cent per annum. Diabetic eye disease is the greatest single cause of blindness in people under 60 years of age in the Western world. Laser photocoagulation is an effective treatment for diabetic retinopathy and prevents the development of severe vascular changes which cause visual loss. Population studies have estimated that 30 per cent of diabetic patients have some type of retinopathy and that 10 per cent would benefit from laser treatment.

Morphological changes in the eye are related to the duration of the disease. In type 1 (juvenile onset, insulin dependent) diabetes, 20 per cent of patients have retinopathy after 12 years. This figure rises to 80 per cent after a duration of 20 years. Retinopathy may occur more rapidly in type 2 (mature onset, noninsulin dependent) diabetes. Vascular abnormalities evolve through several stages. In the preclinical stage there is basement membrane thickening, increased vascular permeability, and increased retinal blood flow. In the clinical stage, retinopathy progresses from early background changes to maculopathy and finally to proliferative retinopathy. Proliferative retinopathy may occur without maculopathy and vice versa, but both are present in the late stages of the disease. Patients with type 2 diabetes are prone to develop maculopathy. In early background retinopathy, degenerative changes in blood vessel walls lead to the formation of microaneurysms. Aneurysms later give rise to haemorrhages and leakage of plasma constituents. When excessive leakage occurs, cystoid oedema may be present at the macula. Alternatively, patients may develop capillary closure with subsequent new vessel formation on the surface of the retina or at the optic disc. This is the proliferative retinopathy stage. Haemorrhage from new blood vessels growing forward from the retina causes the formation of fibrous bands in the vitreous and results in retinal detachment. Untreated eyes with new blood vessels at the optic disc have a 40 per cent risk of severe visual loss within 2 years.

Although vascular lesions can be observed ophthalmoscopically, fluorescein angiography is essential for identifying the stage of retinopathy and deciding when laser treatment is desirable. It is not possible to provide this investigation for all diabetic patients at every medical review and a means of patient selection would be extremely useful. Visual acuity is

Table 9.4a Visual function and capillary non-perfusion at the macula for five diabetic patients having proliferative retinopathy with disc new vessels

Patient	Age	Duration of diabetes in years	Visual acuity	Macular assessment	Goldmann visual field loss (arbitrary units) 30°	Total
1	39	17	6/5	Good. Fine capillary drop out. Small areas of focal capillary leakage	52	64
2	44	21	6/6	Mild background retinopathy. Small areas of focal capillary leakage	50	148
3	49	36	6/6	Mild capillary drop out. Moderate oedema	110	912
4	42	16	6/12	Ischemic macula considerable leakage and capillary non-perfusion	158	620
5	47	5	6/9	Marked maculopathy. Ischemia temporal to the fovea and considerable leakage at the posterior pole.	172	1208

The numerical values for normal fields are 2078 at 30° and 4956 for total field.

not a helpful indicator of retinopathy, and selection in terms of duration of diabetes is not entirely satisfactory because patients may have been unaware of the disease for a number of years. The detection of acquired colour deficiency therefore has an important role in selecting patients, at risk of visual impairment, who would benefit from a full ophthalmic investigation.

Results obtained with a battery of colour vision tests show that the severity of colour deficiency corresponds with visual field loss and with the extent of macular involvement (Table 9.4a,b). The results are typical of the progression of type 3 defects. In the initial stages of background retinopathy, hue discrimination and threshold short wavelength sensitivity (measured with the TNO test) confirm the presence of a slight type 3 defect. As the retinopathy develops further, a type 3 defect is clearly demonstrated by pseudoisochromatic tests and by grading tests such as the D15 and desaturated D15 tests. In severe proliferative retinopathy and maculopathy, the patient is functionally tritanopic but all three colour vision mechanisms are affected and red–green errors as well as tritan errors are made on clinical tests. Adaptation tests, such as the TNO test, confirm that the short-wave mechanism is absent (acquired tritanopia) and the F–M 100 hue test shows poor overall hue discrimination confirming that all three colour vision mechanisms are involved (Fig. 9.1).

A monocular examination with a full test battery takes a long time to

Table 9.4b Results for a colour vision test battery for five diabetic patients with acquired type 3 defects graded by degree of capillary non-perfusion at the macula

Patients listed in rank order of macular involvement	Ishihara test number of mistakes	AO (HRR) test diagnostic tritan plates	Farnsworth D15 test	F–M 100 hue test error score	TNO test threshold blue sensitivity at 460 nm	Isochromatic lines
1	None	Pass	Pass	132	60%	Slight tritan
2	4	Pass	Pass	128	7%	Slight tritan
3	7	Pass	Fail	288	None	Moderate tritan
4	17	Fail	Fail	396	None	Tritanopic
5	24	Fail	Fail	500+	None	Tritanopic

Table 9.5 Percentage of diabetic patients failing the Ishihara plates and the new City University tritan plates according to the severity of background retinopathy

Diabetic patients	Ishihara plates	New City University tritan plates Screening	New City University tritan plates Grading
Without retinopathy	0	18	5
With retinopathy	15	53	20
Degree of retinopathy			
Mild	1	21	0
Moderate	10	54	18
Severe	20	70	41

complete and is not practical for screening. Suitably chosen paired tests can give reliable information and take only a few minutes to administer. A combination of the Ishihara plates and the new City University plates is effective for the detection of moderate or severe type 3 defects which are found in severe preproliferative retinopathy, proliferative retinopathy, and maculopathy (Table 9.5). Three results are possible.

1. Failure of the small colour difference tritan screening designs is found in about a third of patients with background diabetic retinopathy. This result is a more reliable indication of retinopathy in young patients than in patients over 50 years of age. However a slight type 3 defect suggests that more severe retinal changes may develop rapidly and that retinal status should be monitored regularly.

2. Failure of both the screening and diagnostic tritan designs suggests severe preproliferative or proliferative retinopathy and the patient should be referred to an ophthalmologist for a detailed retinal assessment.

3. Patients who fail both the tritan plates and the Ishihara plates should be referred as a matter of priority. About 80 per cent of patients with severe preproliferative retinopathy and proliferative retinopathy obtain this result and laser treatment may need to be considered. Similar results coupled with failure of the Amsler chart indicate the presence of maculopathy. Colour vision is extremely poor if cystoid macular oedema is present.

The SPP 2 plates can be used to detect diabetic retinopathy but a combination of tritan and red–green responses is always obtained as a consequence of the test design. This means that the SPP 2 test is less useful for monitoring the stages of the type 3 defects.

Many diabetic patients are over 50 years of age and screening for type 3 defects is influenced by physiological changes in the eye media. A number of false positive tritan results on clinical tests is inevitable in older people and these results have to be balanced against the overall benefit that colour vision screening achieves. Tests based on computer controlled displays have achieved some success in predicting the onset of severe diabetic retinopathy.

Laser photocoagulation, given at an appropriate stage of diabetic eye disease, prevents further vascular changes taking place and preserves visual acuity. Treatment styles vary according to the characteristics of the retinopathy. Panretinal photocoagulation is used to treat proliferative or severe preproliferative retinopathy. Treatment involves placing a large number of laser burns in the peripheral retina and between the vascular arcades. The amount of treatment depends on the severity of the retinopathy and 2000–3000 burns are usually needed. A small number of burns are applied to individual vascular malformations near the macula or, in the case of diffuse abnormalities, a grid pattern of about 150 burns is placed centrally.

Alteration in colour vision may be found following photocoagulation treatment. The amount of change depends on the wavelengths of the laser used, the number of burns and the exposure time. Panretinal photocoagulation of more than 2000 long duration (0.5 s) argon laser burns has been found to produce permanent acquired tritanopia and field defects. A characteristic tritan axis of confusion is obtained with the F–M 100 hue test in these cases (Fig. 9.2a). Treatment with long duration burns has been discontinued. Photocoagulation with short duration (0.05 s) burns preserves the neural retinal structure and has little effect on the visual field. This treatment produces a temporary reduction in short-wave sensitivity and a slight type 3 deficiency which recovers after a few days. For example, results for paired D15 test show a type 3 (tritan) defect 1 week after treatment and recovery 6 weeks afterwards (Fig. 9.6). The success of the treatment in controlling vascular abnormalities can be assessed by monitoring this recovery. The mechanism which produces acquired

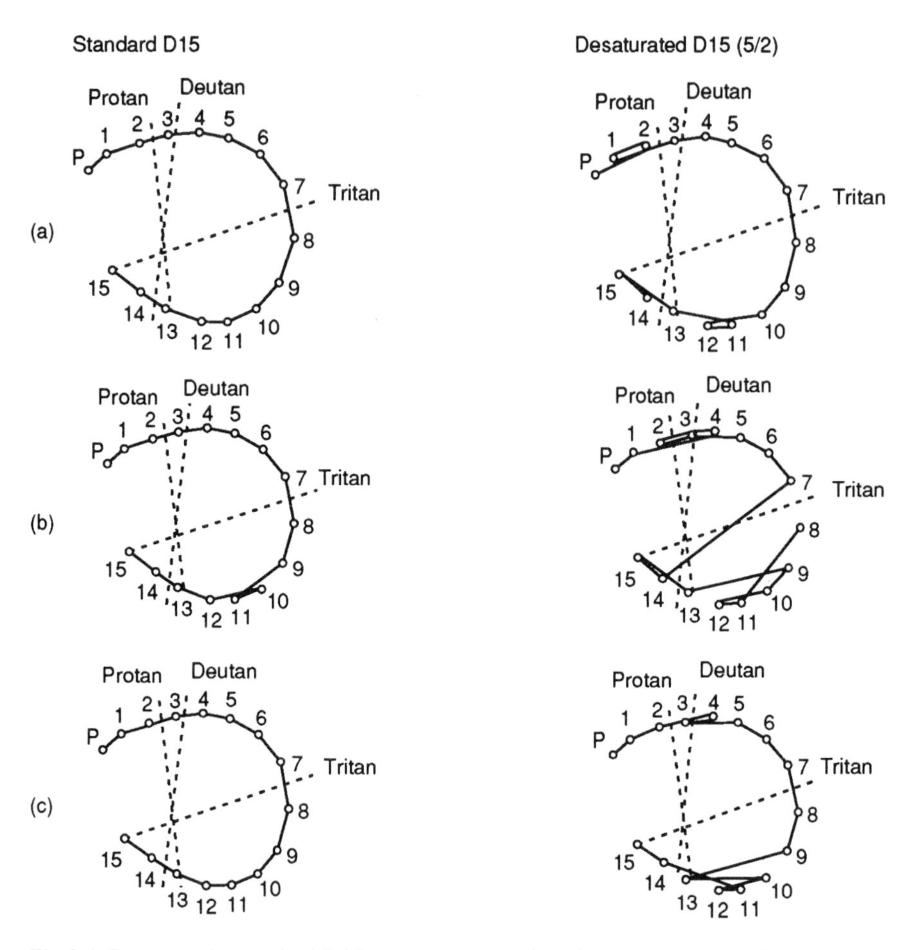

Fig 9.6 Errors on the standard D15 test as a means of confirming the F–M 100 hue result in monitoring the effects of photocoagulation treatment in diabetic retinopathy. Results are shown for the standard D15 and for a desaturated test with Munsell 5/2 before treatment (a), 1 week after treatment (b), and 6 weeks after treatment (c). Before treatment the F–M 100 hue error score was 128 with a tritan axis of confusion. At 1 week after treatment with 3000 short duration (0.05 s) argon burns, the error score was 168 and a more marked tritan axis of confusion was present. Six weeks after treatment, the error score was 80 and no axis of confusion could be discerned.

colour deficiency following photocoagulation is thought to be phototoxic damage due to intraocular light scatter during treatment. Such damage is dose related and wavelength specific. Colour vision changes are reduced if the blue (488 nm) wavelength is filtered from the argon laser, or if lasers with wavelengths longer than 520 nm are used.

Diabetic patients with severe congenital or acquired colour deficiency may be unable to monitor their own glycosuria or blood glucose levels using colour coded tests. Patients who have both congenital and acquired colour deficiency are particularly at risk. Mistakes of two steps in interpretation of the Ames Clinitest may result in poor control of blood glucose levels. Congenital red–green dichromats who acquire a type 3 deficiency have extremely poor colour vision and may become monochromatic. Colour deficient patients should ask another family member to confirm the results of these colour coded tests.

(c) Lesions of the optic nerve

Acute optic neuritis occurs most frequently in people between 20 and 40 years of age. Visual loss is monocular and progressive. Vision either deteriorates very suddenly or changes gradually over about 2 weeks. Colour vision is severely affected leading to a type 2 red–green defect and there is a parallel loss of the central visual field. Recovery may take 4–6 weeks. Although 90 per cent of patients recover normal visual acuity, slight colour deficiency may remain. A unilateral deutan-like colour vision defect therefore suggests that a previous episode of optic neuritis has taken place.

The visual deficit in optic neuritis is due to demyelination and axonal loss in the optic nerve. Permanent atrophic damage to the optic nerve can be detected by electrodiagnostic tests after each episode. Although optic neuritis has a variety of causes, and rare inherited forms of the disease have been described, the most frequent cause is multiple sclerosis. Multiple sclerosis patients may suffer periodic attacks of optic neuritis, which increase in frequency and severity, leading to permanent loss of vision. Patients describe the initial symptoms as a central 'gauze' or coloured mist. Later, all colours appear to be desaturated and a variety of colours are confused including reds and greens. Recovery of colour deficiency can be monitored using the F–M 100 hue test (Fig. 9.2b). An overall loss of hue discrimination resolves into a red–green defect resembling a congenital deutan deficiency as visual acuity improves.

Type 2 acquired defects are found in patients who develop toxic optic neuritis due to excessive use of tobacco or exposure to industrial chemicals such as lead and thallium.

A typical type 3 tritan defect occurs in autosomal dominant optic atrophy. In this relatively benign disease of unknown aetiology, atrophy is thought to result from oedema of the optic nerve rather than demylination. Severe colour deficiency, resembling tritanopia, is found when visual acuity is only minimally affected and disc pallor is barely noticeable ophthalmoscopically. The discovery of tritan defects in this condition led to the suggestion that tritanopia might always be due to this cause and that the tritanopes found in the *Picture Post* survey were in the early stages of the disease. The autosomal dominant mode of inheritance tended to support this view. The controversy was resolved by recalling the *Picture Post* tritanopes for another examination 20 years after the first. Patients with optic atrophy would have developed visual field defects and have very poor visual acuity after this interval. No losses in visual function were found, confirming that congenital tritanopia exists as a separate entity.

(d) Acquired colour deficiency due to intracranial lesions

Type 3 (tritan) defects are reported in lesions of the chiasma and optic radiations but there is no clear association with specific field defects.

Type 3 defects are sometimes found in unilateral posterior hemisphere lesions, which cause hemianopic field defects, but there is no consensus as to whether these are found after injury to the right or to the left hemisphere or both. Isolation of colour disturbance within the affected portion of the visual field has been reported in some cases.

Complete colour blindness (cerebral achromatopsia) may result from localized vascular lesions which involve the ventromedial occipitotemporal cortex. Colour disturbance can occur very rapidly and, in some cases, prior to other symptoms. The patient may wake up in the morning and find that everything appears in 'black and white' or 'in shades of grey'. Occasionally loss of colour perception is incomplete and the person complains that all colours look extremely pale. Visual acuity remains normal but there is bilateral upper visual field loss, prosopagnosia and impaired topographical memory. In addition, the patient may be aphasic and alexic. Affected patients are unable to respond to clinical tests and very simple colouring and sorting tasks are substituted. Colour naming and colour association are also examined. For example, the patient may be asked to point to the correct colour reproduction of a familiar object or to colour simple outline drawings by choosing the appropriate colour from a selection of felt-tip pens. The patient has to name the colour of the pen chosen and name the colour associated with individual objects. Some aphasic patients may correctly associate a banana with the colour yellow but be unable to select (or name) the correct colouring pen. Hue discrimination is examined by asking the patient to sort coloured cards into groups. Three cards may be the maximum that the patient can cope with and all he or she needs to do is point to the card which is a different colour from the other two.

Comprehensive examination of patients with incomplete loss of colour perception has been possible in a few cases. Although loss of blue or blue–green vision appears to be a common feature, each case is unique and it is not always possible to classify the defect using the accepted nomenclature. The patient described by Pearlman *et al.* (1979) had sustained a bilateral inferior occipital infarction involving the lingual and medial occipitotemporal gyri bilaterally. This lesion gave rise to bilateral superior hemianopia (with macular sparing), prosopagnosia, and topographical disorientation. Visual acuity was 6/6. The patient described all colours as either pale blue or pink. He was able to pass the Ishihara plates but results for the D15 and F–M 100 hue test showed extremely poor hue discrimination similar to that of a congenital monochromat. Spectral wavelength discrimination measurement found residual sensitivity between 580 and 620 nm only, and a neutral zone (described as grey) occurred in the yellow region of the spectrum between 560 and 570 nm. Wavelength discrimination is therefore similar to that of a congenital tritanope except that the short-wave section of the curve in the violet region of the spectrum is absent. The wavelength of maximum relative luminous efficiency was found to be at 580 nm which is a longer wavelength than that found in deuteranopia. Colour matches with the Nagel anomaloscope were similar to those of a congenital deuteranope. However, isolation of the three chromatic mechanisms, using the Wald–Marré method of selective adaptation, showed absence of the blue mechanism with the red/green mechanisms only minimally reduced. Colour-matching experiments confirmed that the patient was functionally a tritanope.

Other abnormalities of colour vision such as chromatopsia are associated with cortical lesions. Spontaneously generated colour phosphenes can result from a variety of lesions in the visual pathway and cortex. The entire field of view may appear to be suffused in colour and all objects appear covered, for example, in 'gold paint' or 'snow'. The coloured visual patterns associated with migraine are an example of this type of temporary visual hallucination. Occasionally the sensation of colour may be retained for a period when the object is no longer in the field of view. Bright reds sometimes persevere in this way. Additionally, colours may appear to overlap the border of the object or to be separated in a different plain from the object itself. Bizarre effects of this type usually occur in the active stage of intracranial pathology.

Individual patients have been described who show different types of interaction between colour and form vision. Loss of object recognition can occur without affecting colour vision and vice versa. One such patient apparently developed her own colour code by attaching coloured tapes to household objects, such as items of cutlery, which she was otherwise unable to identify. A person with a unique type of congenital colour deficiency has been examined in great detail by Hendricks *et al.* (1981). Although the abnormality is clearly central in origin, no lesion has been detected by NMR scans, suggesting a metabolic cause. The defect involves an inhibitory effect produced by red lights and surfaces. Red is described as producing an achromatic (grey) sensation which spreads beyond the limits of the coloured object and suppresses detection of other stimuli within this zone. Visual acuity and visual fields are normal.

Drug-induced acquired colour vision defects

A large number of therapeutic drugs are reported to produce colour vision disturbance. These reports rely heavily on subjective descriptions and are not supported by test results. However, drugs which are known to produce visual side effects are also likely to affect hue discrimination. Colour vision is changed in proportion to changes in visual acuity. In most cases visual side effects only occur if the normal therapeutic dose is exceeded or if the treatment is unduly prolonged. In a small number of cases, visual disturbance arises from a hypersensitive response following minimal treatment. There are three specific drugs which are well documented as producing acquired colour deficiency and colour vision testing is a useful means of monitoring the level of toxicity.

(a) Digitalis

Digitalis and its derivatives, digoxin and digitoxin, are cardiac glycosides. These drugs are effective in the control of congestive heart failure and certain cardiac arrhythmias. They are widely prescribed. Visual side-effects have been documented for over 200 years. Disturbance of colour vision is a very early sign of toxicity and may occur when plasma concentrations of the drug are considered to be within the normal range. In the initial stages the patient may report scintillating scotomas (including fortification spectra similar to those reported in migraine), episodes of coloured vision, and transient visual field defects. Objects may take on a 'snowy' appearance or appear covered in a reddish haze (erythropsia). In severe cases of drug toxicity, patients suffer from symptoms resembling optic neuritis, including loss of visual acuity, a central scotoma and a type 2 acquired red–green defect. In these cases there is ample documentary evidence, shown by changes in the F–M 100 hue test results, that the severity of colour deficiency corresponds with the plasma concentration of the drug. Colour vision recovers after the drug is withdrawn.

(b) Ethambutol

Ethambutol is prescribed in the treatment of tuberculosis. Toxic side-effects can occur either after therapy with high dosage or after prolonged therapy at low dosage. Doses of less than 25 mg/day are rarely toxic but it appears that toxicity is more likely in patients who have associated alcoholism or diabetes. Symptoms are typical of optic neuritis and are of sudden onset. A type 2 acquired red–green colour deficiency occurs. The drug is slowly eliminated from the body and colour vision improves gradually after withdrawal. In severe cases colour vision may be permanently impaired.

(c) Chloroquine and thioridazine

Chloroquine is used in the treatment of connective tissue disorders, such as rheumatoid arthritis and systemic lupus erythematosus, and for malaria prophylaxis and therapy. Chloroquine is a melanotropic compound. The drug has an affinity for the retinal pigment epithelium and may be stored in the retina for several years even after treatment is discontinued. Visual impairment is due to the cumulative dose and occurs as a result of prolonged therapy. Very few cases of chloroquine toxicity have been reported where the cumulative dosage is less than 300 g. This is equivalent to a daily dosage of 250 mg for 3 years. It is important to establish an early diagnosis of retinal chloroquine toxicity since withdrawal of the drug is insufficient to prevent further damage. Colour vision tests can be used to monitor chloroquine therapy and increasingly frequent periodic colour vision examinations are recommended if the drug is given for a number of years. A type 3 acquired tritan defect occurs initially and this develops into a type 1 red–green defect in the later stages. Episodes of chromatopsia have also been reported. Loss of hue discrimination is particularly severe if a central pigmentary retinopathy develops.

Pigmentary retinopathy may occur following treatment with thioridazine. This drug is prescribed for some psychiatric disorders and is also melanotropic. In advanced cases, large areas of retinal pigment epithelium atrophy develop. A type 3 acquired colour deficiency has been reported but documentation is incomplete.

Vitamin A deficiency

In Western countries, vitamin A deficiency is associated with alcoholism and with metabolic storage diseases which give rise to chronic liver disorders. The predominant symptoms are night blindness and symmetrical constriction of the visual field. Type 3 acquired tritan defects occur. In severe cases this leads to an overall loss of hue discrimination (Fig 9.2c). Recovery, following aural doses of vitamin A, can be shown by improvement in F–M 100 hue error score.

References

Bronté-Stewart, J. and Foulds, W.S. (1971). Acquired dyschromatopsia in vitamin A deficiency. In *Modern problems in ophthalmology*, Vol. 11 (ed. G. Verriest), 168–73. S. Karger, Basle.

Boyle, R. (1688). Some uncommon observations about vitiated sight. In *Works of Boyle* (ed. T. Birch), 1924, pp. 445–52. John Mollon, London.

Dain, S.J. and Birch, J. (1987). An averaging method for the interpretation of the Farnsworth–Munsell 100 hue test. *British Journal of Physiological Optics*, 7, 3, 267–80.

von Helmholtz, H. (1897). *Physiological optics*. Trans. J.P.C. Southall. Optical Society of America, New York, 1924.

Hendricks, I.M., Holliday, I.E., and Ruddock, K.H. (1981). A new class of visual defect. *Brain*, **104**, 813–40.

Köllner, H. (1912). *Die Störgungen des Farbensinnes. Ihre Klinische Bedeutuna und Ihre Diagnose*. Karger, Berlin.

Konig, A. (1897). *Über Blaublindkheit*. Sitzung Akademie Wiss, pp. 718–31. Berlin.

Marré, M. (1973). The investigation of acquired colour deficiencies. In *Colour 1973*, pp. 99–135. A. Hilger, London.

Marré, M., Marré, E., Zenker, H.-J., and Fulle, D. (1989). Colour vision in a family with autosomal cone dystrophy. In *Colour Deficiencies IX*. pp. 181–7. Kluwer, Dordrecht.

Pearlmann, A.L., Birch, J., and Meadows, J.C. (1979). Cerebral colour blindness. *Annals of Neurology*, **5**, 253–61.

Wilson, G. (1855). Researches on colour blindness, Sutherland and Knox, Edinburgh.

Further reading

Foster, D.H. (ed.) (1991). *Inherited and acquired colour vision deficiencies*. MacMillan, London.

Marré, M. and Marré, E. (1986). *Erworbene Storungen des Farbensehens*. Georg Thieme, Leipzig.

Rietbrock, N. and Woodcock, B.G. (ed.) (1983). *Color vision in clinical pharmacology*. Vieweg and Son, Braunschweig.

10. Filter aids for colour deficient people

During the Second World War many young men looked for ways of correcting their colour deficiency so that they could become pilots. Many 'cures' were offered at the time including electrical stimulation of the eye, injections of iodine, staggering doses of vitamins, flashing light therapy, and tuition in colour naming. The situation reached such proportions in the US that the American Society of Ophthalmology and The Association of Schools and Colleges of Optometry issued a statement that there was no 'cure' for colour deficiency and that coaching men to pass colour vision screening tests might endanger national security and decrease efficiency in industry.

It is not possible to 'correct' colour deficiency either with filters or by any other means. However, looking through coloured filters may help people to distinguish isochromatic colour pairs. In the nineteenth century both Seebeck and Maxwell experimented with possible filter aids to assist colour deficient people. Seebeck suggested looking through a coloured liquid and Maxwell invented a pair of spectacles which were divided horizontally into two parts, one red and one green. Spectacle lenses divided vertically into red, clear, and green portions were also tried. Looking at a scene with and without a filter, or alternately through different filters, helps the person to see luminance contrast differences between areas of colour which initially look the same. For example if someone cannot distinguish between red and green objects, looking through a red filter will make the green object look much darker than the red one. The light–dark relationship is then reversed when looking through a green filter. With practice it is possible to learn these contrast relationships and give the correct colour names to the objects seen. Multicoloured spectacle lenses look rather bizarre but enable the wearer to make rapid colour comparisons by glancing through different sections of the lens. Most people would not want to wear such a device in public but there are sufficient advantages to make the arrangement acceptable for recreational activities, such as stamp collecting. Filters may also be hand-held or mounted in a frame which can be pivoted into place when necessary.

The X-Chrom lens

Placing a red filter in front of one eye was first suggested by Mauthner in 1894 and the idea was popularized by Cornsweet as a 'cure for colour blindness' in his textbook on visual perception published in 1970. In this case any improvement in colour discrimination is achieved by monocular comparison of the colours seen (Taylor 1982). A clinical device in the form of a monocular contact lens, the X-Chrom lens, was introduced by Zeltzer in 1971 and has been fitted to a number of colour deficient people in the US. The X-Chrom lens is a dyed PMMA (polymethylmethacrylate) corneal contact lens and is fitted to the nondominant eye. It is also possible to dye soft contact lens materials but the dye may be impermanent in high water content lenses. The material shows maximum transmission above 575 nm with some additional transmission below 480 nm. The overall transmission depends on the thickness of the worked lens but the colour is remarkably dense even in a thin lens. The designers of the lens claim that it provides significant enhancement of colour perception for red–green colour deficient observers. The claim is endorsed by satisfied wearers of the lens but the results of detailed investigations do not confirm that this is so. Most reports which support the X-Chrom lens do so on the grounds that performance on pseudoisochromatic tests is improved and that some colour deficient people are able to obtain the correct result when wearing the lens. These observations have led to the false assumption that a 'correction' has been achieved. Pseudoisochromatic plates are carefully designed to have a precise colour appearance when illuminated with white light. Looking through a filter changes both colour appearance and luminance contrast and this enables colour deficient people to obtain the correct result. The effect of wearing a filter is similar to illuminating pseudoisochromatic plates with coloured light; both of these viewing conditions make the design unsuitable for its original function.

The aim of the X-Chrom lens is to enhance general hue discrimination and it is more appropriate to evaluate the lens by looking for improved performance on hue discrimination tests, such as the DI5 test and the F–M 100 hue test, than on pseudoisochromatic plates. Normal trichromats make more mistakes with the F–M 100 hue test when looking through a coloured filter and the axis of confusion obtained by colour deficient observers rotates. Rotation of the axis of confusion shows that although improvements in hue discrimination are found in some parts of the hue circle, deterioration occurs in parts where hue discrimination was previously good. A similar alteration is found with the D15 test and the diagnosis of the type of colour deficiency may change when wearing the lens (Fig. 10.1). An example of alteration in the F–M 100 hue plot with and without the X-Chrom lens worn in front of one eye is shown in Fig. 10.2.

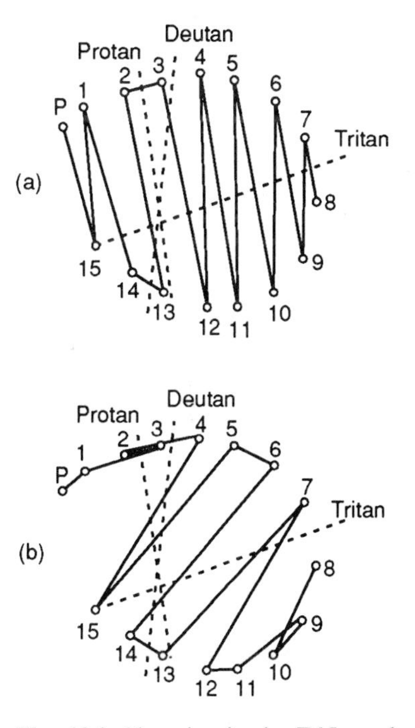

Fig. 10.1 Alteration in the D15 result for a protanope viewing through a red filter. Results are shown for a protanope with the standard D15 test without a filter (a) and when viewing through a red filter (b). Alteration in isochromatic colour confusions are produced. When looking through the filter the result is similar to that found in rod monochromatism.

Two colour-deficient people, a severe protanomalous trichromat and a deuteranope, have been examined with a complete test battery with and without the X-Chrom lens (Paulson 1980). Both observers had been wearing the lens long enough to overcome any initial adaptation problems. The results are summarized in Table 10.1. The protanomalous observer failed the same colour vision tests both with and without the lens. Fewer

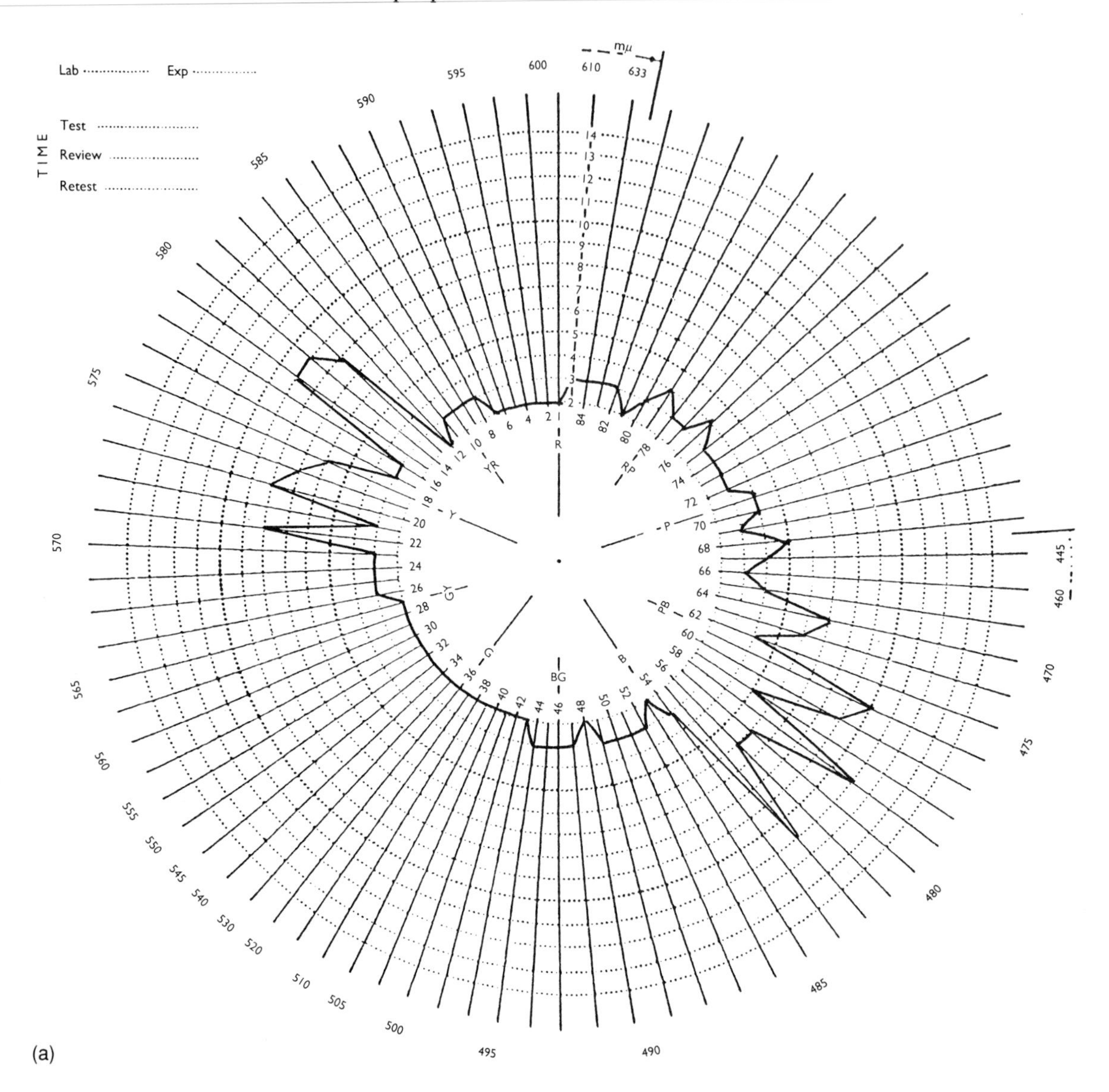

Fig. 10.2 Alteration in the F–M 100 hue result for a deuteranomalous trichromat viewing through the X-Chrom lens. The results for a deuteranomalous trichromat, with a Nagel matching range of 0–30 scale units, show a deutan axis of confusion and an error score of 144 (a). When viewing through the X-Chrom lens, the error score is reduced to 72 but poor hue discrimination extends over a larger area of the purple quadrant of the hue circle than before (b).

errors were made on pseudoisochromatic tests when wearing the lens, but his performance on the D15 and H16 tests was poorer and an enlarged matching range was obtained on the Hecht–Schlaer anomaloscope. The Hecht–Schlaer anomaloscope is similar in design to the Nagel anomaloscope. There was no significant difference in the F–M 100 hue error score or on the error score obtained with the Farnsworth lantern. The diagnosis of the type of deficiency changed from protan to deutan with the HRR plates and the D15 test.

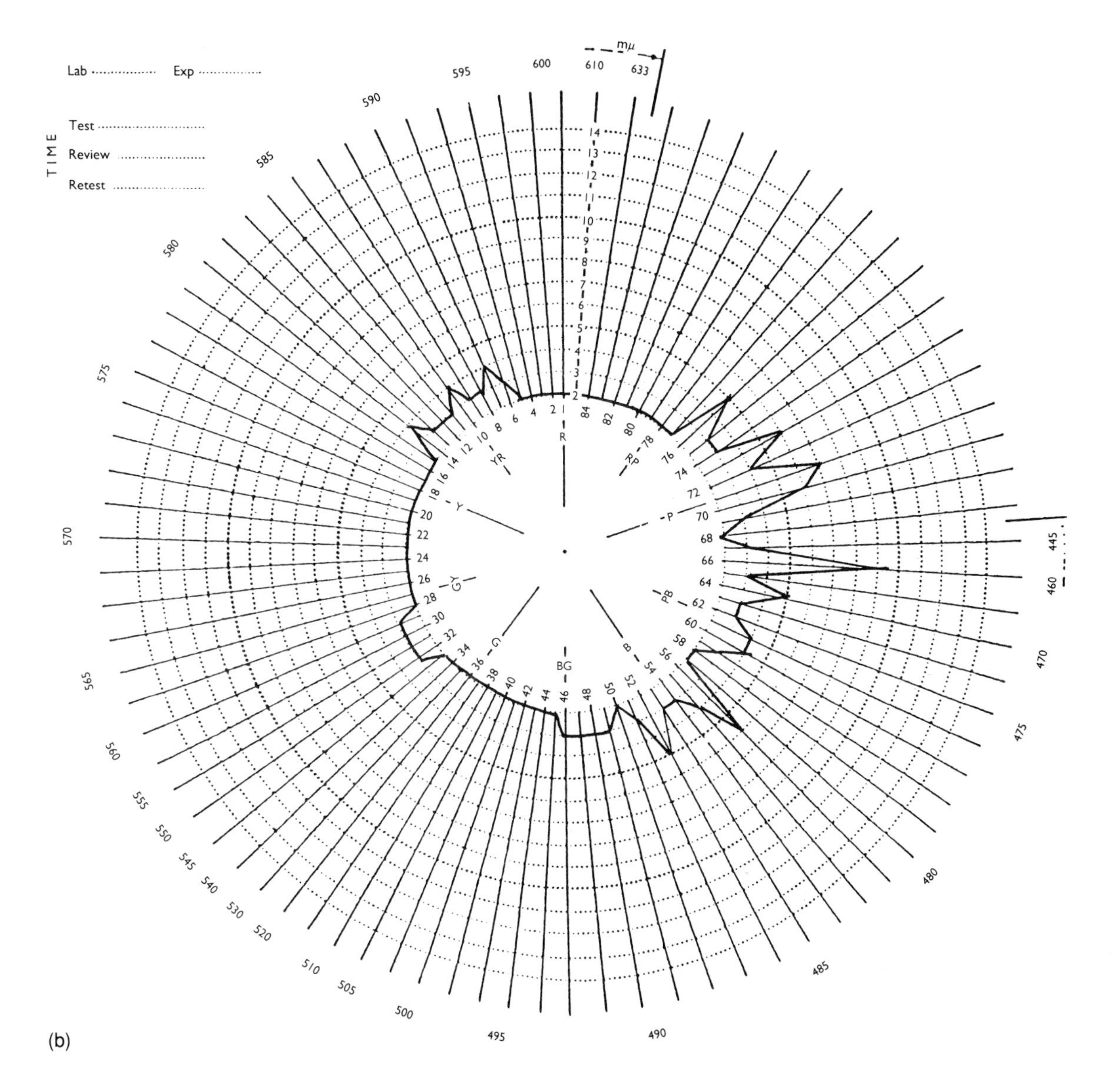

(b)

The performance of the deuteranopic observer improved on some tests when wearing the lens but not on others. He was able to pass the Ishihara test, the D15, and H16 tests with the lens, but there was no change in the error score obtained with the Farnsworth lantern and performance on the F–M 100 hue test was significantly poorer. The F–M 100 hue test and the Farnsworth lantern are tests of practical colour recognition ability and these results do not provide evidence of enhanced colour vision in either of these two X-Chrom lens wearers although both were highly motivated to do well.

Wearing a monocular filter produces other changes in visual function. These may include poorer visual acuity in the eye wearing the lens, tilting of the image plain caused by the Pulfrich effect and poor stereoscopic

Table 10.1 Colour vision test results for a deuteranope and a protanomalous trichomat with and without the X-Chrom lens

Test	Deuteranope		Protanomalous trichromat	
	Without lens	With X-Chrom lens	Without lens	With X-Chrom lens
Ishihara plates number of errors	14	2	16	11
AO (HRR) plates diagnosis	Strong deutan	Mild red–green	Medium protan	Medium deutan
Farnsworth lantern (Falant) error score	6.5	5.5	7	7.5
Farnsworth D15 number of crossings	12	1	2	4
H16 number of crossings	10	1	0	3
F–M 100 hue error score	256	316	184	168
Hecht–Schlaer anomaloscope matching range	Full	Full	20–70	10–80

From Paulson 1980.

vision. Many wearers report a lustre or glittering appearance of red objects which is derived from retinal rivalry. The long term effect of wearing the lens has not been evaluated.

Although filters may not assist overall hue discrimination they can be of considerable help with individual colour recognition problems. For example, protanopic doctors and optometrists have difficulty distinguishing patches of black melanin pigment and small haemorrhages in the retina. A red filter can be inserted into the aperture wheel of an ophthalmoscope, or slit lamp, so that the retinal appearance can be compared with and without the filter. Haemorrhages 'disappear' when the filter is in place but melanin pigment still looks black against the red background. Other pathological changes such as blanching of vascular tissue can be assessed in the same way. Filters can be placed in microscopes to help individual users to distinguish differentially stained areas of prepared specimens or into other optical instruments, such as endoscopes, to aid clinical diagnosis. If the type of colour deficiency is known it is possible to select an appropriate filter to enhance the difference between any isochromatic colour pair.

Coloured filters have recently been considered as a device for improving reading ability in dyslexic children. The possible effect of these filters on both normal and abnormal hue discrimination should not be overlooked.

References

Cornsweet, T.N. (1970). *Visual perception*, pp. 194–8. Academic Press, New York.

Mauthner, L. (1894). *Farbenlenhre*, pp. 146–7. J.F. Bergman, Weisbaden.

Paulson, H. (1980). The X–Chrom lens for the correction of colour vision. *Military Medicine*. **145**, 8, 557–60.

Taylor, S.P. (1982). The X Chrom lens—a case study. *Ophthalmic and Physiological Optics*, **2**, 165–70.

Zeltzer, H.I. (1971). The X Chrom lens. *Journal of the American Optometric Association*, **42**, 933–9.

Further reading

Schmidt, I. (1976). Visual aids for correction of red–green colour deficiencies. *Canadian Journal of Optometry*, **38**, 38–47.

11. The occupational consequences of defective colour vision

The possible dangers of employing colour deficient train drivers were recognized by Wilson in the mid-nineteenth century and the first colour vision tests were designed to screen railway workers. Railway recruitment boards throughout the world continue to screen applicants and colour deficient people are not employed as drivers, signalmen, maintenance engineers, or lineside workers. Similar recruitment policies apply to maritime transport and to aviation. Colour deficient people are not employed as pilots, engineers, vehicle drivers, deck officers, or signals personnel in the armed forces.

Normal colour vision is required in occupations which make extensive use of connotative colour coding and in occupations where failure to use a colour code correctly might be a safety hazard (Table 11.1a). The requirement for train drivers to have normal colour vision is readily accepted but the enforcement of colour vision standards for pilots is often questioned by disappointed applicants, especially in the US and Australia where personal air transport is often the best way of travelling long distances. Colour codes are used for in-flight information as well as for navigational aids and the public demand the highest standard of safety in transport systems where a single error might lead to great loss of life or destruction of property. The same high standards are demanded in electrical and electronics industries, the police force, and the fire service. Mistakes with electrical colour codes might cause a serious accident and incorrect descriptions of clothing or vehicles might prevent police officers from identifying suspects or obtaining convictions in criminal cases. Firemen need to recognize colour coded materials, such as inflammable chemicals, in poor viewing conditions. Normal colour vision is needed for occupations involving colour quality assurance, colour matching, and colour reproduction. In consequence, colour deficient people may not be employed in the paint and textile industries, the motor industry, or in colour printing and fine art reproduction. Decisions made by colour

Table 11.1a Careers and occupations requiring normal colour vision

The Armed Services
Officers in the Navy, Air Force, Army, and Marine service; pilots, engineers, and vehicle drivers
Merchant navy officers and seamen
Customs and excise officers
Civil aviation
Airline pilots, engineers, airport technical and maintenance staff, air traffic controllers
Railways
Train drivers, engineers, and maintenance staff
Electrical and electronic engineers
Hospital laboratory technicians and pharmacists
Police and fire service officers
Workers in paint, paper, and textile manufacture—or in photography and fine art reproduction
Workers in industrial colour quality assurance

deficient people in key posts in these industries might lead to expensive commercial mistakes. Good colour vision may also be required in stock control, retailing colour goods, or advising on colour schemes for interior decoration.

Colour vision standards apply to personnel doing a specific task and it is not usually necessary for everyone in a particular industry to have normal colour vision. However, managerial staff must have normal colour vision if they supervise colour quality assurance. Training programmes may also influence colour vision requirements. For example, in the UK, railway management recruits undergo a period of training alongside drivers and all Naval officers perform the duties of an officer of the watch at some stage in their career. The training structure therefore demands that all recruits have normal colour vision even though colour is not involved in the final occupation. Different countries have specific colour vision requirements. For example, Malaysian bank employees need to have normal colour vision because currency notes are different colours but the same size.

Occupations in which defective colour vision is a disadvantage are listed in Table 11.1b. In these occupations, the need for normal colour vision depends on the importance of the colour task, the frequency with which colour judgements have to be made, and the availability of additional clues to aid colour recognition. In some jobs it is sufficient to recommend that colour deficient employees take care with colour tasks and seek the advice of a colleague if difficulties arise. However, some employers decide that colour judgements are important enough to demand normal colour vision and others are prepared to accept recruits with slight colour

Table 11.1b Careers and occupations in which colour deficiency is a handicap

Bacteriology
Botany
Cartography
Chemistry
Interior design
Histopathology
Horticulture
Geology and metallurgy

deficiency only. This leads to a lack of a uniform policy which makes it difficult to offer precise careers advice. For example, some local transport authorities require drivers of public service vehicles to have normal colour vision and others do not. The problem then arises that a colour deficient person employed by one local authority may find himself refused transfer to another authority later in his career. This sort of difficulty needs to be anticipated and should be explained to the individual beforehand.

Most dichromats and 75 per cent of anomalous trichromats are aware that they have colour vision problems. Some people react by denying that this is so and others consider that their problems are trivial. As a result a number of colour deficient people attend job interviews knowing that colour deficiency is a disadvantage. In Steward and Cole's (1989) study, almost 25 per cent of the people questioned reported that they had been denied employment because of colour deficiency. Some people thought it 'worth a try' but others were extremely surprised and angry to be turned away.

Everyday colour vision problems

Most colour deficient people are likely to minimize their colour vision problems at a job interview and more accurate information is obtained from informal discussions and confidential surveys. The survey made by Steward and Cole (1989) found that over 75 per cent of colour deficient people have difficulty with everyday tasks. They also confirmed that the type of error and the frequency that mistakes are made is clearly related to the severity of colour deficiency as shown by the Nagel anomaloscope matching range. Dichromats have more problems than anomalous trichromats but even slight anomalous trichromats make mistakes. Typical problems include selecting coloured goods and judging the ripeness of fruit such as tomatoes or apples. Most people have difficulty picking ripe cherries or strawberries and cannot identify plants with similar shaped flowers. Many colour deficient people make mistakes matching sewing threads and in buying home decorating materials. Choosing similar brown furniture stains causes problems as well as selecting paints and wallpapers.

Some colour deficient people cannot see when roast meat is cooked or when young children become sunburned. Pallor, as a sign of ill health, occasionally goes unnoticed and adjustment of the colour quality of the family television set is a source of disagreement between family members. Most people with severe colour deficiency ask for help buying clothes or in selecting appropriate dress for special occasions. Some find it necessary to label garments with colour names or to store particular colours in the same place. Distinguishing black and brown garments is specially difficult.

Colour deficient people report difficulty with colour coded maps, such as the British Ordnance Survey series, and have to rely more on written information. Red and black lines to denote footpaths and bridleways cannot be distinguished by protans. Most geographical maps use grey or orange for urban areas, green for woodland, and green–yellow–red codes to denote height above sea level and yellow–orange–red codes for roads; all these colours are potentially difficult for colour deficient people to distinguish unless lightness differences can be detected. However blues or purples are easily seen on this background and are effective for showing important routes such as motorways or flight paths. Colour deficient people often need twice as long as normal trichromats to extract information from these maps. Some city transport maps have elaborate colour codes. In this case it is the number of different colours which causes problems. For example the London Underground map has twelve lines which are all shown in different colours and inevitably some colours look the same to colour deficient people.

In Steward and Cole's survey nearly a quarter of the colour deficient people questioned reported difficulty with sporting activities. Playing or watching snooker causes the most problems and even slight anomalous trichromats find the brown ball difficult to identify. Some recreational sailors cannot distinguish coloured light signals and marker buoys. This problem is well known and professional mariners must have normal colour vision, but, as yet, there is no corresponding standard for leisure activities. In contrast, both professional and recreational pilots are subject to similar colour vision standards. Some ball-game players report difficulty distinguishing team colours. Although some colours are easy to distinguish at the beginning of the game, this can become increasingly more difficult if clothing becomes wet or dirty. One soccer referee described how he relied on his wife to place his red and yellow cards in different pockets before the game so that he could always produce the correct card when the situation required.

Congenital tritan defects are rare and little attention has been given to occupational difficulties which might arise from this type of colour deficiency. There are problems distinguishing dark 'navy' blue and black garments and difficulty distinguishing yellow/violet and white/black colour pairs.

Problems driving vehicles

Cole and Steward found that 50 per cent of dichromats and 20 per cent of anomalous trichromats have difficulty distinguishing road traffic signals.

Traffic signals are the only connotative colour code which most people use regularly. Information is provided by both the position and sequence of the lights and it is surprising that so many colour deficient people say that they have problems. Difficulties are not confined to particular lighting conditions or times of day but are reported in normal daylight, bright sunlight, and at night. Confusion of street lights with traffic lights is described by about a third of colour deficient people, many of whom feel insecure driving at night because of the large number of overhead lights. Colour distortion produced by sodium street lighting generally makes matters worse and small sodium lights can be mistaken for red stop signals. People with protan defects are particularly at risk when driving at night because red lights are difficult to see. Many protanopes describe driving past red stop signals at night but often attributed this to inattention rather than poor colour vision. The sequence of paired traffic lights is different in some countries and this makes it more difficult for visiting drivers to anticipate the colour of the next signal. Other reported driving problems include failure to distinguish red and green 'cat's-eyes', used as lane markers on motorways, and poor visibility of directional information on a brown background such as road signs used in the UK to locate places of historic interest.

Evidence that protans have particular driving problems is found from road accident statistics collected in Europe (Neubauer *et al.* 1978). People with protan defects are involved in some types of accident, such as rear-end collisions and failure to halt at stop signals, much more frequently than other drivers. However the same statistics show that colour deficient drivers are involved in fewer accidents of all types. A possible explanation is that colour deficient people learn to take extra care. Experiments, which simulate driving conditions, show that all colour deficient people need to be closer to red signal lights than normal trichromats before the colour is recognized. The recognition distance is reduced by as much as 50 per cent for protans and the breaking distance approaching a stop signal is halved. These data support recommendations that drivers of public service vehicles, such as ambulances and buses, should have normal colour vision. Some countries also ban people with protan defects from driving heavy goods vehicles and, in the past, both Austria and Romania have banned protans from driving private cars at night. Car driving is an essential part of modern life and most people would not support imposing driving restrictions on colour deficient people. A more acceptable alternative would be to change the traffic signalling system to make it easy for everyone to interpret. Improved recognition could be obtained by illuminating intersections where there are traffic lights, increasing the intensity of the red light at night, or by giving signals different shapes. For example, a red 'stop' signal could be shown as a cross and a green 'go' signal as an arrow. Shaped signals are already used at a number of intersections, but changing all traffic lights to this system would be very costly.

Problems with colour codes

Most colour deficient people have difficulty interpreting colour codes containing reds, yellows, and greens. Even codes consisting of three colours can be difficult to distinguish if no luminance contrast exists. For example, the three-way colour code for domestic electrical wiring in the UK was changed in the 1960s from red, black, and green, denoting 'live', 'neutral' and 'earth', to brown, blue, and green/yellow stripe to make it safe for colour deficient people to use. The equivalent code in the US is black, white, and blue–green. Codes which include black and white are not always favoured if there is a possibility of the white surface getting dirty. A three-way code consisting of white, cyan (turquoise) and orange is easy for colour deficient people to distinguish.

High quality electrical components with numerical codes are now being manufactured, but most resistors and cables are still colour coded. The colours of electrical codes have different chroma and value which enables luminance contrast to be used to assist identification. Even so, the study by Voke (1976) showed that on average, dichromats misname 20 per cent of resistor colours and anomalous trichromats misname 10 per cent. People who failed the D15 test or obtained an error score of over 150 on the F–M 100 hue test made more mistakes than people who performed well on these tests. Both protans and deutans confuse the following colours:

grey and violet
green and brown
green and red
red and brown
violet and blue
orange and yellow.

Similar mistakes were made naming and matching coloured wires in multicore cables. The most common mistakes were matching blue with violet, red with brown, red with green, and orange with green. When the cables were denoted with twisted colour pairs the most difficult pairs to distinguish were red/blue and brown/violet twisted stripes. The consequences of confusing electrical wires and resistors depends on the size of the difference step denoted by the code. In resistors this depends on which band is misread. For example, a 'small error' in identifying a colour band which indicates a multiplication factor is potentially much more dangerous than a 'small error' in measuring resistance.

Similar colour-naming mistakes are made with other colour coded tasks. Many examples can be found in schools. Colour codes containing ten colours are used for teaching arithmetic and colour is used to highlight particular words in reading books. Colour codes are used to distinguish drugs and chemical substances and changes in colour must be recognized to denote the end point in some chemical reactions. Errors equal to two colour difference steps in codes used to assess glycosuria can lead to poor management of diabetic patients. Colour deficient doctors and patients need to take special care interpreting these codes and, if necessary, enlist the help of an assistant or relative to check their results.

Large areas of colour are always easier to identify than small ones. The colours used to distinguish pipelines, chemicals, and gas cylinders are relatively large and additional written information is provided to supplement the colour code. In industry, chemicals and gas cylinders are stored in particular places and workers are able to take their time making an identification. However, speed of recognition in unfamiliar surroundings is needed in an emergency and it is important that police officers and firemen have normal colour vision.

Colour codes have to be used safely in different lighting conditions. Changes in illumination can alter colour appearance and either make colour recognition easier or more difficult for colour deficient workers. These changes cannot be predicted and some personnel officers reject all colour deficient applicants, even when hue discrimination is good in standard viewing conditions rather than risk difficulties in other types of illumination.

Elaborate colour codes are used in some computer systems. Visual display units do not cause problems if colour deficient people are able to select their own colour schemes or adjust the luminance contrast of the colours to suit their individual requirements. This may not be possible if the same colour code has to be used by a large number of people. For example, fixed colour codes are sometimes used for public service information, details of manufacturing processes, and corporate business systems. Red lettering on a black background should not be used to display information because it cannot be seen by protans. Red and black lettering on a white background, once favoured by bank managers to distinguish in-balance and overdrawn accounts, is also impossible for protans to distinguish.

Careers advice

Some careers are barred to colour deficient people for clearly defined safety reasons. In addition it is possible to compile a long list of occupations which involve colour and might cause problems for colour deficient people. However, most colour deficient people can perform a range of colour tasks successfully. Giving occupational advice therefore consists of estimating the error risk and the possible consequences if mistakes are made. In some occupations the complexity of colour coded work increases on promotion and this has to be taken into account when employing recruits. If a decision is made to employ a colour deficient person he or she must be advised about possible future limitations and that transfer to another company, doing similar work, may not be possible if a different policy is adopted by the second company.

Ideally each person's colour recognition ability should be assessed in relation to the requirements of a particular job. However a detailed 'job-by-job' analysis would be extremely time-consuming. The studies by Steward and Cole (1989) and Voke (1976) have clearly shown that practical ability is related to performance on clinical tests and advice is usually given on the results obtained with a test battery. This enables the type of colour vision to be determined and standardized examination procedures ensure that similar results are always obtained.

The initial step is to establish whether a person's colour vision is normal or abnormal using a screening test. If abnormal colour vision is found, estimating the severity of colour deficiency is the second, and most important, step. Diagnosis of whether a person is a dichromat or anomalous trichromat is not always needed but it is essential to classify protan and deutan defects. Protanopes are likely to be the most severely handicapped in performing practical tasks, followed in rank order by deuteranopes and protanomalous trichromats. Deuteranomalous trichromats are the least handicapped and slight deuteranomalous trichromats are minimally affected.

The needs of a particular occupation are categorized using the following criteria:

1. The occupation requires normal colour vision and excellent colour matching ability. A screening test is used and all colour deficient people are excluded from employment. People with normal colour vision are given an additional test, such as the F–M 100 hue test, to assess practical hue discrimination ability. A pass/fail level can be based on an acceptable error score relevant to the occupation.
2. The occupation requires normal colour vision. All colour deficient applicants are identified and excluded from employment by failure of a screening test.
3. The occupation is suitable for people with slight colour deficiency. The exclusion of people with 'significant' colour deficiency is the most difficult aspect of giving colour vision advice. Grading tests, such as the Farnsworth D15 test and the City University test, have been specifically designed for this function but it may be preferable to give a full battery of colour vision tests to determine the type as well as the severity of colour deficiency. Four categories of colour deficiency can be distinguished from the results obtained (Table 11.2).
4. Colour deficiency is a minor handicap and no colour vision standard is necessary. A screening test is used to identify colour deficient recruits. These people are advised to take extra care with colour coded tasks and to seek the advice of a colleague if difficulties arise.

A full test battery should include more than one test for estimating the severity of colour deficiency. However in some instances two tests will suffice. The Ishihara plates are used to provide accurate red–green screening, and the second test is used to estimate the severity of colour deficiency and identify tritan defects. The second test can be either the HRR plates (if available), the D15 test, or the City University test. A lantern test is often given as the second test but the severity of colour deficiency cannot be determined from the results. It is preferable to give a recognized grading test, or a complete test battery, before a lantern test is given so that the type of colour deficiency is known. Protans and severe deutans can then be excluded from further examination and the lantern test given to slight deuteranomalous trichromats only. A similar procedure is recommended for a 'resistor wire' test in the electrical industry. The procedure for distinguishing people with normal colour vision, slight, or severe colour deficiency and for selecting people for colour quality assurance is shown in Table 11.3.

Table 11.2 Categories of colour deficiency estimated from performance on grading and diagnostic tests as a guide to suitability for occupations involving colour judgements

Colour vision category	General recommendations	Colour recognition	Farnsworth D15	TCU test	F–M 100 hue test	Nagel anomaloscope Diagnosis	Nagel anomaloscope Matching range
1	Suitable for all types of work except colour matching and colour quality control	Normal	Pass	Pass	Average error score. No axis of confusion	DA	Small
2	Suitable for most tasks except recognition of very pale or very dark colours and when speed is not important	Slightly impaired	Pass	Pass	Below average error score. Poor discrimination or slight diagnostic axis of confusion	DA	Small to moderate
3	Suitable for tasks involving recognition of large colour difference only. Errors occur with bright colours in some viewing conditions	Moderately impaired	Fail with partial errors	Fail with four errors or less	Moderate error score. Diagnostic axis confusion	DA PA	Moderate Moderate
4	Not suitable for tasks involving colour recognition. Errors occur in all viewing conditions	Severely impaired	Fail with complete errors	Fail with five errors or more	High error score. Marked diagnostic axis of confusion	DA PA D P	Large Large Complete Complete

Superior colour vision

Excellent hue discrimination is needed for work in industrial colour quality assurance. Superior hue discrimination is an acquired skill which is developed through training and by observing an experienced person making colour judgements. Most people with normal colour vision can acquire this skill if they are sufficiently motivated to do so. Careful observation is needed and not everyone has the personality necessary to achieve success. Possible colour aptitude can be estimated from initial performance on a hue discrimination task. The F–M 100 hue test is usually selected for this

Table 11.3 Initial colour vision assessment for quality assurance

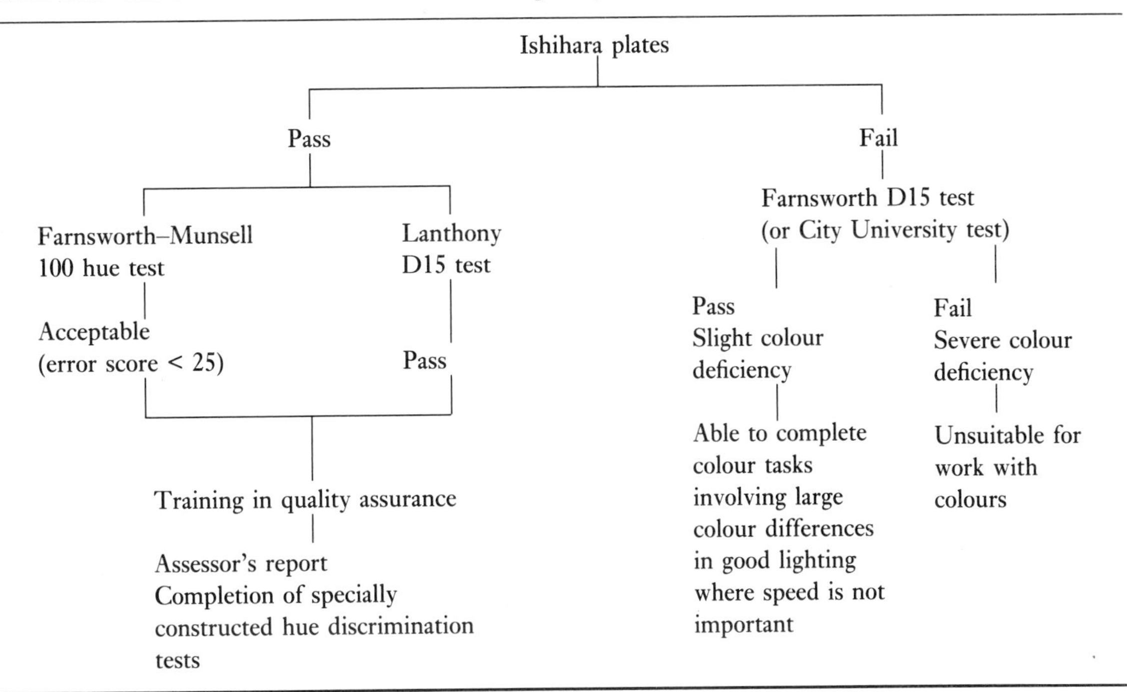

and an acceptable error score set, for example an error score of less than 25 allows the person to make 6 single placement errors. The Lanthony D15 test might also be used. If these tests are completed successfully, the recruit begins work alongside a skilled colour quality assessor who discusses and demonstrates acceptable and unacceptable colour matches by reference to colour tolerance sets. The length of training varies according to the occupation and the range of colours involved, but is unlikely to be less than two weeks. At the end of the training period the assessor reports whether the necessary accuracy has been achieved. A further objective test may be desirable at this stage but needs to be individually designed and constructed. Ideally the test consists of the same paints, manufactured items, or components that are being produced and is given in the job location under the same illumination as that normally available. Munsell samples can be considered as an alternative to manufactured items. Hue discrimination improves for large areas of colour and when colours are placed edge to edge rather than separated. The Farnsworth–Munsell tests all display colours in small caps and the colours are separated by the width of the cap rim. A different format is needed for industrial tests so that appropriately-sized colour samples can be placed in juxtaposition. The recruit either places the samples in sequence according to a particular attribute or states whether a displayed colour pair is an acceptable match. It is essential that the test is formalized and the results recorded. Colour matching skills need to be exercised regularly and a period of retraining may be necessary if a person is away from the job for any length of time.

References

Cole, B.L. and McDonald, W.A. (1988). Defective colour vision can impede information acquisition from redundantly colour-coded video displays. *Ophthalmic and Physiological Optics*, **8**, 188–210.

Neubauer,O., Harrer, S., Marré, M., and Verriest, G. (1978). Colour vision and traffic. In *Modern problems in ophthalmology*, Vol. 19 (ed. G. Verriest), pp. 77–81. Karger, Basel.

Steward, J.M. and Cole, B.L. (1989). What do colour defectives say about everyday tasks? *Optometry and Visual Science*, **66**, 5, 288–95.

Vingrys, A.J. and Cole, B.L. (1988). Are colour vision standards justified in the transport industry? *Ophthalmic and Physiological Optics*, 8, 257–74.

Voke, J. (1976). The industrial consequences of deficiencies of colour vision. Ph.D. thesis. The City University, London.

Index